CRITICAL THINKING & LOGICAL REASONING WORKBOOK-4

4

GIFT OF LOGIC™ SERIES

Boost Your Thinking Skills

An Essential Resource for Everyone

Verbal Reasoning

Analytical Reasoning

Pictorial Reasoning

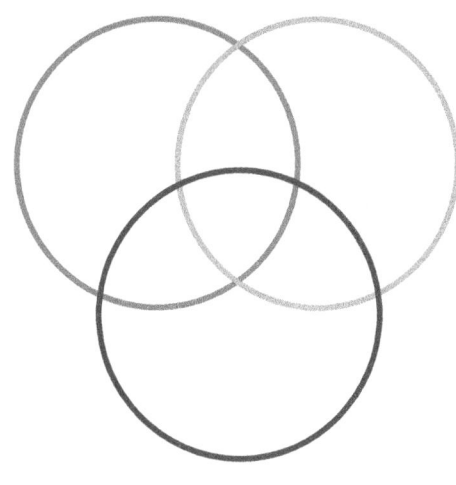

THIRD EDITION

| FOR GRADES 3-5 | STUDENTS, TEACHERS, AND PARENTS |

Ranga Raghuram

GIFT OF LOGIC™

Gift Of Logic, Inc

http://www.giftoflogic.com
sales@giftoflogic.com

Critical Thinking and Logical Reasoning Workbook-4
ISBN-13: 978-1494832308
ISBN-10: 1494832305

Third Edition
1-2014

Copyright © 2009 Gift Of Logic, Inc. All rights reserved. No part of this publication may be reproduced, stored in a retrieval system, transmitted in any form or by any means, electronic, mechanical, photocopying, recording or otherwise, without the written permission of the publisher.

License: This book is licensed for use by one person only. Use of this book in a group setting (classroom, workshop, etc) without the written permission of the publisher is prohibited. Unauthorized duplication is strictly prohibited by law. Contact the publisher at sales@giftoflogic.com for classroom/school/group licensing.

GIFT OF LOGIC™
CRITICAL THINKING & LOGICAL REASONING CURRICULUM
12 WORKBOOKS TO BOOST YOUR THINKING SKILLS

For Kindergarten, Grade 1, and Grade 2

Workbook# 0

Verbal Reasoning	Finding the truth, Inferencing, Analogies, Synonyms and Antonyms, Agree/Disagree
Analytic Reasoning	Memory drill, Decision making, Positioning, Sudoku
Pictorial Reasoning	Connect the dots, Mazes, Picture Sequence, Spot the difference, etc

Workbook# 1

Verbal Reasoning	Finding the truth, Inferencing, Analogies, Synonyms and Antonyms, Agree/Disagree
Analytic Reasoning	Sorting, Positioning, Picking, Assorted problems, Numeric and Alphabetic Sudoku
Pictorial Reasoning	Picture Sequence, Spot the difference, Odd picture

Workbook# 2

Verbal Reasoning	Finding the truth, Classification, Direct and Inverse relationship, Inferencing, Analogies, Agree/Disagree
Analytic Reasoning	Sequencing, Scheduling, Strategy, Picking, etc
Pictorial Reasoning	Picture Analogy, Odd picture, Pattern matching, etc

For Grade 3, Grade 4, and Grade 5

Workbook# 3

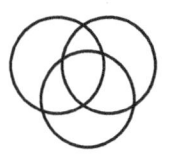

Verbal Reasoning	Not, And, Or, If .. then, Conditional inferencing, Unconditional inferencing, Symbolic Logic
Analytic Reasoning	Lists, Sequencing, Grouping, Venn Diagrams, Graph logic, Number logic, Letter logic, Sudoku
Pictorial Reasoning	Picture sequence, Picture analogy, Odd picture, Picture difference, Pattern matching

Workbook# 4

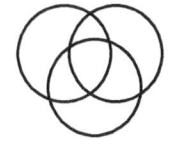

Verbal Reasoning	Contradiction, Converse, Inverse, Contrapositive, Conditional inferencing, Symbolic Logic
Analytic Reasoning	Scheduling, Looping, FIFO, LIFO, Correlation, Venn Diagram, Graph logic, Number logic, Sudoku, etc
Pictorial Reasoning	Picture sequence, Picture analogy, Odd picture, Picture difference, Pattern matching

Workbook# 5

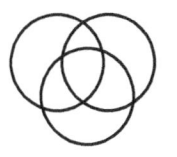

Verbal Reasoning	Biconditional, Categorical inferencing, Cause and Effect, Symbolic Logic, Agree/Disagree, Word and Sentence analogy
Analytic Reasoning	Correlation, Grouping, Venn Diagrams, Graph logic, Number logic, Letter logic, Sudoku, etc
Pictorial Reasoning	Picture sequence, Picture analogy, Odd picture, Picture difference, Pattern matching

********* Essential resource for everyone *********
*http://www.giftoflogic.com   *sales@giftoflogic.com

GIFT OF LOGIC™
CRITICAL THINKING & LOGICAL REASONING CURRICULUM
12 WORKBOOKS TO BOOST YOUR THINKING SKILLS

For Grades 6-12, College/University Students, Adults

Primer

Prereq

Verbal Reasoning	Logical Operators, Conditional, Categorical and Causal reasoning, Validity, Fallacies, Symbolic Logic
Analytic Reasoning	Positioning, Grouping, Sudoku
Pictorial Reasoning	Pattern perception, Figure formation, Paper folding and cutting, Figure matrix, Rule detection

Workbook# 6

Verbal Reasoning	Arguments-Main point, Must be true, Cannot be true
Analytic Reasoning	Positioning, Grouping, Sudoku
Pictorial Reasoning	Pattern perception, Figure formation, Paper folding and cutting, Figure matrix, Rule detection

Workbook# 7

Verbal Reasoning	Arguments-Strengthening, Weakening
Analytic Reasoning	Positioning, Grouping, Sudoku
Pictorial Reasoning	Pattern perception, Figure formation, Paper folding and cutting, Figure matrix, Rule detection

Workbook# 8

Verbal Reasoning	Arguments - Controversy, Paradox
Analytic Reasoning	Positioning, Grouping, Sudoku
Pictorial Reasoning	Pattern perception, Figure formation, Paper folding and cutting, Figure matrix, Rule detection

Workbook# 9

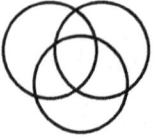

Verbal Reasoning	Arguments- Assumptions, Reasoning strategy
Analytic Reasoning	Positioning, Grouping, Sudoku
Pictorial Reasoning	Pattern perception, Figure formation, Paper folding and cutting, Figure matrix, Rule detection

Workbook# 10

Verbal Reasoning	Arguments-Flawed reasoning, Analogous reasoning
Analytic Reasoning	Positioning, Grouping, Sudoku
Pictorial Reasoning	Pattern perception, Figure formation, Paper folding and cutting, Figure matrix, Rule detection

********* Essential resource for everyone *********
Get the GIFT OF LOGIC™ today !
*http://www.giftoflogic.com *sales@giftoflogic.com

Dear Reader:

Your decision to purchase this book is commendable. You now have in your hands, a comprehensive, easy-to-read book in Critical thinking and Logical reasoning that will introduce you to three different areas of thinking and reasoning - Verbal, Analytical and Pictorial. Solving problems in Verbal Reasoning is important to develop a critical mind. Solving problems in Analytic Reasoning is important to develop a flexible and resourceful mind. Solving problems in Pictorial Reasoning is important to develop a visually alert mind.

This book is presented in a workbook format to help you progress quickly. Parents and teachers are urged to complete the exercises ahead of the student and assist them whenever necessary with the help of detailed answers provided at the end of the book. This book can be used as a supplementary resource in the regular class room or it can be used during winter and summer vacations. College/University students, working professionals and retired individuals will also find the Gift Of Logic(tm) Series very useful in enhancing their problem solving abilities, confidence and general intellect.

Critical thinking and Logical reasoning must be practiced consistently to develop strong cognitive skills. After completing the exercises in this book, continue to read the other books in this series to get familiar with different types of Logical reasoning problems.

This workbook is one in a series of twelve workbooks. Please refer to the brochure before this page for a brief description of each workbook. Visit the website http://www.giftoflogic.com for more information.

Happy thinking and reasoning!

TABLE OF CONTENTS

Verbal Reasoning

Converse statements 9

Inverse statements 13

Contrapositive statements 15

Contradictory statements 17

Redundant statements 20

Inferencing
    conditional - must be true 21
    conditional - cannot be true 25

Agree 28

Disagree 29

Sentence Analogy 30

Word Analogy 32

TABLE OF CONTENTS

Analytic Reasoning

List Processing	35
Sequencing	38
Scheduling	39
Looping	40
FIFO and LIFO	42
Correlation	44
Grouping	45
Venn Diagram	49
Graph Logic	52
Number Logic	60
Letter Logic	63
Sudoku	66

Pictorial Reasoning

Picture Sequence	73
Picture Analogy	76
Odd Picture	79
Picture Difference	81
Pattern Matching	83

Answers

Verbal	86
Analytic	104
Pictorial	137

Certificate of Completion

VERBAL REASONING

Name _____ Date _____

CONVERSE STATEMENTS - conditional

Conditional statements are normally represented by the "If P then Q" format, as discussed in workbook#3.

The converse of a conditional statement "If P then Q" is "If Q then P". That is, the converse of a conditional statement is obtained by reversing the antecedent and the consequent. Converse statements are also called reverse statements or vice versa statements.

    Conditional statement: If P then Q ($P \rightarrow Q$)
    Converse statement: If Q then P ($Q \rightarrow P$)

If a conditional statement is true, then its converse is false. That is, the converse of a conditional cannot be inferred (invalid inference).

Example :

Conditional: If Pam watches a movie, then Quayle will also watch.
Converse: If Quayle watches a movie, then Pam will also watch.

If the conditional is true, then the converse cannot be inferred. If Pam watches a movie, then Quayle will also watch. But, we cannot infer that if Quayle watches a movie, then Pam will also watch. It is possible that Quayle watches a movie with someone else other than Pam.

    Conditional: Pam watches movie $\rightarrow$ Quayle watches movie
    Converse: Quayle watches movie $\rightarrow$ Pam watches movie

The converse statement cannot be inferred from the conditional statement.

Name —————————————— Date ——————————————

CONVERSE STATEMENTS - conditional

Answer the following questions and explain your reasoning.

1

Conditional: If a movie is good, people will watch it.
Inference: If people watch a movie, then it must be a good movie.
Is the inference valid?   A) Yes   B) No
Reasoning:

2

Conditional: If it is 10 PM, the theater will close.
Converse: If the theater closes, then it is 10 PM.
Is the converse a valid inference? A) Yes   B) No
Reasoning:

3

Conditional: Snake bites → will hurt
Converse: hurts → snake bite
Is the converse a valid inference?   A) Yes   B) No
Reasoning:

4

Conditional: very warm → plants will die
Converse: plants will die → very warm
Is the converse a valid inference? A) Yes   B) No
Reasoning:

Verbal Reasoning   Answers-86
© Gift Of Logic, Inc * Copying prohibited

Name _____   Date _____

CONVERSE STATEMENTS - unconditional

In statements that are not conditional, the converse may be true or false depending on the relationship between the two subjects. You need to find this on a case by case basis.

Example 1:
    Statement: Margie weighs more than Rachel.
    Converse: Rachel weighs more than Margie.

In the example above, if the statement is true, the converse statement is clearly false. Normally, if an unconditional statement is true, the converse is false and cannot be inferred (invalid inference). But, this is not always the case. In some cases, if an unconditional statement is true, the converse can also be true and can be inferred (valid inference).

Example 2:
    Statement: The head is attached to the neck.
    Converse: The neck is attached to the head.
In this example, the statement is true, and its converse can be inferred.

Example 3:
    Statement: The red circle is completely inside the blue circle.
    Converse: The blue circle is completely inside the red circle.
    The converse is a   A) Valid inference   B) Invalid inference

This clearly an invalid inference. This can be verified by drawing the two circles, one inside the other and verifying that the inference is false.

Verbal Reasoning
© Gift Of Logic, Inc * Copying prohibited

Name _____   Date _____

CONVERSE STATEMENTS - unconditional

Write the converse of the given statements and evaluate whether it is true or false.

---

1 Statement: All books are study materials.
Converse:
The converse is a   A) Valid inference   B) Invalid inference

---

2 Statement: The red string is tied to the blue string.
Converse:
The converse is a   A) Valid inference   B) Invalid inference

---

3 Statement: The boat race was held after the swimming race.
Converse:
The converse is a   A) Valid inference   B) Invalid inference

---

4 Statement: Abdul met Ali.
Converse:
The converse is a   A) Valid inference   B) Invalid inference

---

5 Statement: Kathy and Cindy are friends.
Converse:
The converse is a   A) Valid inference   B) Invalid inference

Name _____   Date_____

INVERSE STATEMENTS - conditional

The Inverse of a conditional statement is obtained by <u>negating</u> both the antecedent and the consequent. Consider the following example:

    Conditional: If the bus arrives, we can go home.
    Inverse: If the bus does not arrive, we cannot go home.

In the above inverse statement, if the bus does not arrive, we can still go home by catching a taxi or by taking a ride from a friend. So, the inverse statement is not a valid inference.

The inverse can be represented in symbolic form as follows:
        Conditional: $P \rightarrow Q$
        Inverse: $\sim P \rightarrow \sim Q$

Negation using the NOT symbol ~ was discussed in workbook# 3. If a conditional statement is true, then its inverse is false. That is, the inverse of a conditional cannot be inferred (invalid inference).

Questions 1-4 that follow test your understanding of the inverse statements. Write the inverse statement and symbolic forms.

1

    Conditional: If the building is old, it will be painted.
    Inverse:
    Conditional Symbolic:
    Inverse Symbolic:

Is the inverse true or false? A) True   B) False
Reasoning:

Name _____   Date_____

INVERSE STATEMENTS - conditional

2

Statement 1: If the pencil is sharp, the handwriting will be good.
Statement 2: The pencil is not sharp.
Statement 3: The handwriting will not be good.

Does statement 1 and statement 2 together imply statement 3?
    A) Yes   B) No
Reasoning:

---

3   Conditional: The alarm will sound if there is a fire.
    Inverse:
    Conditional Symbolic:
    Inverse Symbolic:

Is the inverse true or false? A) True   B) False

Reasoning:

---

4   Conditional: If the soil is not fertile, the plants will not grow.
    Inverse:
    Conditional Symbolic:
    Inverse Symbolic:

Is the inverse true or false? A) True   B) False
Reasoning:

Verbal Reasoning   Answers-88
© Gift Of Logic, Inc * Copying prohibited

Name _____   Date _____

CONTRAPOSITIVE STATEMENTS - conditional

The contrapositive of a conditional statement is obtained by <u>switching and negating</u> the antecedent and consequent parts of the conditional statement.

    Conditional: If P then Q
    Contrapositive: If not Q then not P

Inference: If a conditional statement is true, then its contrapositive is also true. Therefore, the contrapositive is a valid inference.

<u>Symbolic:</u>
    Conditional: $P \rightarrow Q$
    Contrapositive: $\sim Q \rightarrow \sim P$

<u>Example:</u>
    Conditional: If it rains, the roads will be slippery.
    Contrapositive: If the roads are not slippery, then it did not rain.

<u>Symbolic:</u>
    Conditional: rains $\rightarrow$ slippery
    Contrapositive: $\sim$slippery $\rightarrow$ $\sim$rain

If the conditional statement is true, then the contrapositive statement can be inferred. In this example, if it rains, the roads will be slippery and by contrapositive inference, if the roads are not slippery, then we can conclude that it did not rain.

Name _____   Date _____

CONTRAPOSITIVE STATEMENTS - conditional

Write the contrapositive statements for the following conditionals.

1 Conditional: If there is inflation, prices will go up.
Contrapositive:

Symbolic:
   Conditional:
   Contrapositive:
Is the contrapositive a valid inference? A) Yes   B) No

---

2 Conditional: If it is foggy, then the airplane cannot land.
Contrapositive:

Symbolic:
   Conditional:
   Contrapositive:
Is the contrapositive a valid inference? A) Yes   B) No

---

3 Conditional: If the weather is good, the rocket will take off.
Contrapositive:

Symbolic:
   Conditional:
   Contrapositive:
Is the contrapositive a valid inference? A) Yes   B) No

Verbal Reasoning   Answers-90
© Gift Of Logic, Inc * Copying prohibited

Name _____   Date _____

CONTRADICTORY STATEMENTS - conditional

The contradiction of a conditional statement "If P then Q" is "If P then Not Q" or "If Not P then Q". Contradictory statements are inconsistent statements.

Conditional: If P then Q; $P \rightarrow Q$
Contradiction: If P then Not Q: $P \rightarrow \sim Q$
Contradiction: If Not P then Q: $\sim P \rightarrow Q$

If a conditional statement is true, then the contradiction is an invalid inference.

Conditional Statement: If I win a lottery, I will become rich.
Contradiction: If I win a lottery, I will not become rich.

Note that "if" the conditional statement is true, then the contradictory statement is false and therefore cannot be inferred.

---

1 Conditional: If you want dinner, you must come home.
Symbolic:
Contradiction:
Symbolic:
Contradiction is A) True   B) False

---

2 Conditional: If you are talented, then you will be popular.
Symbolic:
Contradiction:
Symbolic:
Contradiction is A) True   B) False

Verbal Reasoning   Answers-91   17
© Gift Of Logic, Inc * Copying prohibited

Name _____   Date _____

CONTRADICTORY STATEMENTS - unconditional

A statement is said to be contradictory or inconsistent if it is logically impossible for it to be true.

Examples:
    Lori is tall and short.
    Mildred is loud and silent.

Several statements together are said to be contradictory if it is logically impossible for all of them to be true at the same time.

Examples of inconsistent set of statements:
    1) No man has landed on Mars.
        Justin has landed on Mars.

    2) If a person is tall, then he is smart.
        John is tall, but not smart.

Note that each set of statements, considered together, cannot be logically possible.

---

1   Make up a contradictory statement:

2   Make up a contradictory set of statements:

CONDITIONAL STATEMENTS - SUMMARY

Conditional statements are normally expressed as "If P then Q".
In symbolic form, P → Q is the same as If P then Q.

The <u>Converse</u> of a conditional statement" If P then Q" is "If Q then P."
A converse statement is also called a reverse statement or a vice-versa.
If the conditional statement is true, its <u>converse is false and cannot be inferred.</u>

The <u>Inverse</u> of a conditional statement "If P then Q" is "If Not P then Not Q". If the conditional statement is true, its <u>inverse statement is false and cannot be inferred.</u>

The <u>Contradiction</u> of a conditional statement" If P then Q" is "If P then Not Q" or "If Not P then Q". If the conditional statement is true, its <u>contradiction is false and cannot be inferred.</u>

The <u>Contrapositive</u> of a conditional statement "If P then Q" is "If Not Q then Not P". If the conditional statement is true, its <u>contrapositive is also true and can be inferred.</u>

So, given a conditional statement, the only inference we can make from it is its contrapositive.

Conditions can also be expressed using words such as Unless and Except. These words can be in the beginning or in the middle of the conditional statements.

Verbal Reasoning
© Gift Of Logic, Inc * Copying prohibited

Name _____   Date _____

REDUNDANT STATEMENTS (Tautologies)

Redundant statements (also called tautologies) are statements that express information by repeating words or by expressing the same information several times using different words.

Examples of redundant statements.

> A bad guy is not a good guy.
> This is this and that is that.
> The glass has nothing in it and is empty.
> He was very fat and obese.
> A law is a law is a law.
> No one is above the law because everyone is below the law.
> Pain is very painful.

It is not a good practice to use redundant statements. They serve no purpose. But, wherever it appears, it is important to be able to recognize it as redundant.

Make up your own redundant statements and write them below. You will immediately understand why you should not use them!

_____

1 _____

2 _____

Verbal Reasoning
© Gift Of Logic, Inc * Copying prohibited

INFERENCING - must be true

1 If the weather is good, the airplane will land.

If the above condition is true, then which one of the following must be true?
 A) If the airplane lands, then we can infer that the weather is good.
 B) If the airplane does not land, we can infer that the weather is not good.

Reasoning:

---

2 If a book sells a lot, it must be a good book.

If the above condition is true, then which one of the following must be true?
 A) If a book sells a lot, it must not be a good book.
 B) If a book is a good book, it will sell a lot.
 C) If a book does not sell a lot, it must not be a good book.
 D) If a book is not a good book, it will not sell a lot.

Reasoning:

| INFERENCING - must be true |

3   When the principal speaks, everyone must listen.

If the above condition is true, then which one of the following inferences can be made?
A) When the principal does not speak, everyone will not listen.
B) If everyone is listening, the principal must be speaking.
C) If everyone is not listening, then the principal must not be speaking.

Reasoning:

4   If apples are grown in a farm, then oranges must also be grown. This year, oranges were not grown in the farm.

From the above condition and information, which one of the following inferences can be drawn?
    A) apples were grown this year.
    B) apples were not grown this year.

Reasoning:

Name _____   Date _____

INFERENCING - must be true

5   A car mechanic found two problems with Jack's car. The problems are such that if the first problem is fixed, then the second problem must also be fixed.

If the above situation is accurately described, then which one of the following is possible?
  A) The mechanic fixed the second problem only.
  B) The mechanic fixed the first problem, but not the second problem.

Reasoning:

6   Amanda must go to the wedding if Bobby also goes. But, Amanda could not attend the wedding due to an emergency.

If the above situation is true, then which one of the following can be concluded?
  A) Bobby attended the wedding.
  B) Bobby did not attend the wedding.

Reasoning:

INFERENCING - must be true

7 If the bus is on time, Jane must not take the train to work. But, on Monday, Jane took the train to work.

From the above information, which one of the following inferences can be made?
   A) The bus was on time
   B) The bus was not on time.

Reasoning:

8 A party was scheduled to begin at 6 PM to honor Jack and Kate for their outstanding achievements. The party must not begin unless Jack and Kate arrives. The party began at 7 PM instead of 6 PM.

From the above information, which one of the following inferences can be made?
   A) Either Kate or Jack or both did not arrive before 7 PM.
   B) Both Kate and Jack arrived before 7 PM.

Reasoning:

Name _____   Date _____

INFERENCING - must be true

9 If Arthur wants to buy a camera, he must buy it from shop A or shop B only. But, Arthur was not able to buy a camera.

From the above information, which one of the following inferences can be made?
 A) Shop A had the camera that Arthur wanted, but shop B did not.
 B) Shop A and shop B both did not have the camera that Arthur wanted.

Reasoning:

INFERENCING - cannot be true

10 If Sue is sick, she must not delay her visit to the doctor.

Which one of the following inferences cannot be true?
   A) If Sue is not delaying her doctor's visit, she must be sick.
   B) If Sue is delaying her doctor's visit, she must not be sick.

Reasoning:

Name _____   Date _____

INFERENCING - cannot be true

11 If you wear a tie, you must be a boy.

If the above condition is true, which one of the following cannot be true?
 A) Nick wore a tie, so he must be a boy.
 B) Nick is a boy, so he must be wearing a tie.

Reasoning:

12 If George wants to play well, he must wear a shoe with cleats.

If the above condition is true, which one of the following cannot be true?
 A) If George is not playing well, he is not wearing a shoe with cleats.
 B) If George is not wearing a shoe with cleats, he is not playing well.

Reasoning:

Verbal Reasoning
© Gift Of Logic, Inc * Copying prohibited

INFERENCING - cannot be true

13 If the food is not fresh, you must not eat it.

If the above condition is true, which one of the following cannot be true?
  A) Food that is not fresh must be eaten.
  B) If you are eating the food, it must be fresh.

Reasoning:

14 You cannot be punctual unless you wear a watch.

If the above condition is true, which one of the following cannot be true?
  A) If you are punctual, you must be wearing a watch.
  B) Even without wearing a watch you can be punctual.

Reasoning:

Name _____   Date _____

AGREE

In this page, someone will be expressing their opinion to you on some subject. You need to reply by agreeing with whatever the person says.

1 We must balance work and play. Too much work or too much play is not going to do us any good.

Agree:

2 This lake is very deep. So, it is better that nobody swims in this lake without a lifeguard being present.

Agree:

3 Water is essential for living. So, sprinkling water in our lawns to keep them green must not be more important than using it for drinking.

Agree:

4 People who litter the streets must be forced to pay a heavy fine.
Agree:

Verbal Reasoning
© Gift Of Logic, Inc * Copying prohibited

Name _____   Date _____

DISAGREE

In this page, someone will be expressing their opinion to you on some subject. You need to reply by disagreeing with whatever the person says.

1 It is okay for visitors to feed zoo animals.
Disagree: _____

2 All students who are in the same class are of the same height.
Disagree: _____

3 In order to live on this earth, everyone must speak two languages.
Disagree: _____

4 Cars are fun to drive. So, we should get rid of motorbikes.
Disagree: _____

5 Even if a movie is bad, we must watch it anyway if others watch it.
Disagree: _____

6 This road is scenic with mountains along the way. So, this road will take us to our destination in the shortest amount of time.
Disagree: _____

Name _____   Date _____

SENTENCE ANALOGY

We sometimes use analogy to explain our ideas.

Read the following statements:
    The traffic in this road is moving as slow as a turtle.

The statement effectively means:
    The traffic in this road is moving very slowly.

It compares the speed of the traffic to the speed of a turtle. Since turtles move very slowly, we can infer that the traffic is moving very slowly.

You must learn to identify the things that are being compared. You must also learn to infer what is actually being said by learning to write the statement without using the analogy.

Example:

Asha behaves like a clown.

What are being compared in the statement?
    Asha's behavior and the behavior of a clown.

Rewrite the statement without using analogy.
    Asha behaves in a funny way.

SENTENCE ANALOGY

1 All the lies that Joe has been telling fell apart like a tower of cards.

What are being compared in the statement?

Rewrite the statement without using analogy.

---

2 The cheeks of the baby feel like cushion.

What are being compared in the statement?

The cheeks of the baby are rough.   A) True   B) False

---

3 Daniel lived like a frog in a well.

What are being compared in the statement?

Daniel travelled to many places.   A) True   B) False

---

4 Victor ran like a tiger.

What are being compared in the statement?

Victor ran   A) very fast   B) very slowly

Name _____ Date _____

WORD ANALOGY

1. Pen is to write as Eraser is to
   A) read   B) clean   C) wash

2. Toyota is to car as Boeing is to
   A) truck   B) airplane   C) bicycles

3. Zoo is to animals as Aquarium is to
   A) chimps   B) fish   C) dogs

4. Airport is to planes as Shipyard is to
   A) water   B) trucks   C) ships

5. USA is to country as Europe is to
   A) country   B) state   C) continent

6. Earth is to Sun as Moon is to
   A) Earth   B) Eclipse   C) Mars

7. Palace is to king as Jail is to
   A) convict   B) clergy   C) cleaner

8. Lion is to forest as Fish is to
   A) house   B) water   C) library

9. Air is to breathe as Water is to
   A) spit   B) throw   C) drink

WORD ANALOGY

10. Car : drive :: Bike :
    A) steer   B) pilot   C) ride

11. Piano : music :: Paint :
    A) sculpture   B) poem   C) drawing

12. State : Governor :: City :
    A) councilman   B) mayor   C) secretary

13. Dog : bark :: Cat :
    A) meow   B) bray   C) neigh

14. Gold : costly :: Dirt :
    A) expensive   B) rare   C) cheap

15. North : south :: East :
    A) east   B) west   C) left

16. Apple : fruit :: Carrot :
    A) root   B) vegetable   C) flower

17. Nurse : doctor :: Mechanic :
    A) surgeon   B) engineer   C) lawyer

18. Body : skeleton :: Building :
    A) frame   B) carpet   C) paint

Verbal Reasoning   Answers-103

ANALYTICAL REASONING

Name —————————————— Date ——————————————

1 LIST PROCESSING - sorting based on one property

The names of five people and their height and weight are shown in the following table.

Name	Height (ft)	Weight (lbs)
Andy	4	30
Bobbie	3	50
Cathy	4	40
Dinesh	3	30
Emma	5	20

Sort and rank the list based on height in descending order and fill the list below.

Name	Rank

Sort and rank the list based on weight in descending order and fill the list below.

Name	Rank

Compare the ranks of each person in the two lists. Why are the ranks different in the two lists?

Analytical Reasoning
© Gift Of Logic, Inc * Copying prohibited

2   LIST PROCESSING - sorting based on two properties

Now, sort the names (from the first table) again, first in descending order of height and if there is a tie, sort on weight in descending order to resolve the tie.

Name	Rank

Sort the names (from the first table), first in descending order of weight and if there is a tie, sort on height in descending order to resolve the tie.

Name	Rank

Analytical Reasoning
© Gift Of Logic, Inc * Copying prohibited

3 LIST PROCESSING - adding to a list

A list is sorted in descending order and currently has the following members in it.

Name	Rank
Zachary	1
Samuel	2
Rudolph	3
Roshan	4
Brandi	5
Anita	6

Rank the list again after adding the following members to the list.
Sandy, Rudy, Brendon, Anderson

Name	Rank
Zachary	1
Sandy	2
Samuel	3
Rudy	4
Rudolph	5
Roshan	6
Brendon	7
Brandi	8
Anita	9
Anderson	10

Are the rankings of all the members the same now?

No.

1 SEQUENCING

February						
Sunday	Monday	Tuesday	Wednesday	Thursday	Friday	Saturday
				1	2	3
4	5	6	7	8	9	10
11	12	13	14	15	16	17
18	19	20	21	22	23	24
25	26	27	28			

Answer the questions below based on the calendar for February.

Raja takes piano lessons every thursday.

1) When was his last class in February?
   A) February 29   B) February 22
2) Which of the following is the thursday after his last class?
   A) February 29   B) March 1

Joshua takes lessons in gymnastics every Monday, Wednesday, and Friday.

1) How many gymnastics lessons did he attend in the first week of February?   A) 3 lessons   B) 2 lessons   C) 1 lesson

2) How many gymnastics lessons did he attend in the last week of February?   A) 3 lessons   B) 2 lessons   C) 1 lesson

3) For how many weeks did he take the lessons three times each week?

4) How many gymnastic lessons did Joshua take during the month of February?
   A) 12   B) 13   C) 15

Analytical Reasoning
© Gift Of Logic, Inc * Copying prohibited

SCHEDULING

same day - another time* another day - same time * another day - another time.

Doctor Harper's Appointment Schedule

	Feb 21	Feb 22	Feb 23
8 AM	Jenny		Arjun
9 AM		Mohan	
10 AM	Asif		Josh
11 AM		Laura	Hana

Answer the following questions based on the appointment schedule.

1) If Jenny wants to change her appointment to the another time on the same day, how many choices does she have?
    A) 1 B) 2 C) 3

2) If Asif wants to change his appointment to another day at the same time, how many choices does he have?
    A) 1 B) 2 C) 3

3) If Mohan wants to change his appointment to another day at another time, how many choices does he have?
    A) 1 B) 2 C) 3

4) If Laura wants to change her appointment to another day at a time that is not 9 AM, how many choices does she have?
    A) 1 B) 2 C) 3

5) Hana wants her appointment to be rescheduled to 8 AM. On which day can she now see the doctor?
    A) Feb 23 B) Feb 22 C) Feb 21

Analytical Reasoning
© Gift Of Logic, Inc * Copying prohibited

Name _____   Date _____

1 LOOPING

Looping means repeating several times. The looping stops when some condition is met. You keep track of loops in a counter.

Write your name in the column marked 'Name'. Every time you write your name, increment (add) the counter by one. Keep writing your name until the counter shows 8 and then stop.

Name	Counter
	1

1) What does the counter keep track of?

2) If you write your name two more times, what number will the counter show?

3) If the counter runs up to 15, how many times would you have written your name?

Analytical Reasoning   Answers-109
© Gift Of Logic, Inc * Copying prohibited

Name _____   Date _____

2 LOOPING

Piggy bank

Christina bought a new piggy bank. Her dad said that he will keep on dropping a dime (10 cents) in it until it adds up to one dollar (100 cents).

Counter	Coins	Total
1	10 cents	10 cents

Fill in the grid - write 10 cents in the 'coins' column every time a dime is dropped into the piggy bank. Keep track of how many times a dime was dropped in the column marked 'Counter'. Keep track of Total as well.

1) When the counter reads 6, how many dimes were dropped?
2) When the 'Total' column reads 40 cents, how many dimes were dropped?
3) When the 'Total' column reads 100 cents, how many times were the dimes dropped?
4) What is the counter reading when 80 cents worth of coins were dropped?
5) Can this counter go beyond 10?   A) Yes   B) No

Analytical Reasoning
© Gift Of Logic, Inc * Copying prohibited

1   FIFO (FIRST IN, FIRST OUT)

First In, First Out is the same as First Come, First Served.

The Genius Elementary School invited a face painter for its Annual Day celebrations. The face painter will paint the faces in different colors and patterns on a first in, first out basis. Any student who wishes to have his face painted can ask the face painter to do so.

Dan, Don and David stood in a line (in that order) waiting for their turn.

1) Who was the first one to have his face painted?
       A) Don   B) Dan   C) David
2) Who was the last one to have his face painted?
       A) Don   B) Dan   C) David
3) The last person in the line is the last one to have his face painted.
       A) True   B) False

Tony first arrived at the doctor's office followed by Andy and then by Gina. The doctor will see them on a first come first served basis.

1) Who did the doctor see first?
       A) Tony   B) Andy   C) Gina
2) Did the doctor see Gina before Andy?
       A) Yes   B) No
3) Who did the doctor see last?
       A) Gina   B) Andy   C) Tony
4) Write the order in which the doctor saw the patients.

2   LIFO (LAST IN, FIRST OUT)

Last In, First Out is the same as First In, Last Out.

People who arrived for a conference were first seated in the first row. When the seats in the first row were filled, they were seated in the second row. This pattern was repeated until the seats in all the rows were filled up. When the conference was over, people in the last row were asked to leave first, followed by the people in the second last row until all the people left the conference. The following table shows where seven people sat during the conference. The fourth row is the last row.

Name	Row
Rick	3
Martin	4
Chuck	3
William	1
Nancy	2
Debby	1
Drew	2

Answer the following questions based on the facts presented above.

1) Martin sat in the last row and was the last one to leave the conference.
    A) True   B) False
2) Nancy left the conference before Rick did.
    A) True   B) False
3) Debby and William left the conference after Nancy and Drew.
    A) True   B) False
4) Who were the last people to leave the conference?

Analytical Reasoning

Name _____   Date _____

CORRELATION

The information that we want is always not readily available all in one place. Correlation is the process of finding the information that we need from different sources and joining them together.

1

Read the information in the two boxes below and answer the questions. Draw a line to show how you are correlating the information.

| chimpanzee   Zoo-A | Zoo-B   10 miles |
| gorilla   Zoo-B | Zoo-A   20 miles |

1) How far should we go to see the gorilla?
2) How far should we go to see the chimpanzee?

---

2   Several students are playing at the school playground. The following information is available. Using this information, answer the questions below.

Jacob 2nd grade	2nd grade - high jump team
Jeff 1st grade	
Jamal 1st grade	1st grade - tennis team
Jindal 2nd grade	

1) In which team can we find Jamal in?
2) In which team can we find Jindal in?
3) Who are all in the high jump team?
4) Who are all in the tennis team?

Analytical Reasoning
© Gift Of Logic, Inc * Copying prohibited

GROUPING

If a group is defined by a specific property, then any member of the group also will have this specific property.

1 All the city buses are painted blue. Bus# 22 is a city bus.

What color is Bus #22?
Why?

2

The Music club requires its members to play at least one instrument. The Drama club requires its members to have acting skills. Calvin can play one instrument, Trevor can play two instruments as well as act, but Josh can neither play an instrument nor act.

Based on the information given above, answer the following questions.

1) Who is eligible to become a member of the Music club?

2) Who is eligible to become a member of the Drama club?

3) Which clubs can Josh become a member of?

Name _____   Date _____

GROUPING

3 The following are the names of the players in the Alpha Sports club and the games that they play.

Randy	soccer, baseball
Will	soccer
Mark	baseball
Gary	soccer, baseball
Zac	soccer
Sean	baseball
Anita	soccer
Britney	baseball
Priti	soccer, baseball
Sidney	soccer

It was decided to divide the club into two groups - one for soccer and one for baseball. Answer the questions below.

1) How many members are there in the sports club?

2) Who are all members of the soccer group and the baseball group? Write their names in the list below.

Soccer Group	Baseball group

3) How many members are there in the soccer group?
4) How many members are there in the baseball group?

Analytical Reasoning Answers-115
© Gift Of Logic, Inc * Copying prohibited

Name ——————————————   Date ——————————————

GROUPING (continued)

4

5) Add up the number of members in the soccer and baseball group. Is this number the same as the number of members in the sports club? Explain.

6) Who are all members of both groups?

Now, fill in the three groups below with the names of players from the Alpha Sports Club.

Soccer Group	Soccer and Baseball Group	Baseball Group

7) How many members are there in the soccer group?

8) How many members are there in the baseball group?

9) How many members are there in the soccer and baseball group?

10) Add up the number of members in the soccer, soccer and baseball, and baseball groups. Is this number the same as the number of members in the sports club?

Analytical Reasoning
© Gift Of Logic, Inc * Copying prohibited

Name _____   Date _____

GROUPING AND SUMMARIZING

5 The residents of a Zoo and their population are shown below.

Residents	Population
Lions	2
Parrots	5
Tigers	3
Eagles	3
Hippopotamus	4
Flamingoes	3

Split the group into birds and animals and write the names of the members of the group and their population in the two boxes shown below.

Birds	Animals

1) How many animals are there in the Zoo?
2) How many birds are in the Zoo?
3) There are more animals in the Zoo than there are birds.
        A) True   B) False
4) More birds must be brought in so that there are equal number of animals and birds in the Zoo.   A) True   B) False

Analytical Reasoning   Answers-117
© Gift Of Logic, Inc * Copying prohibited

VENN DIAGRAM

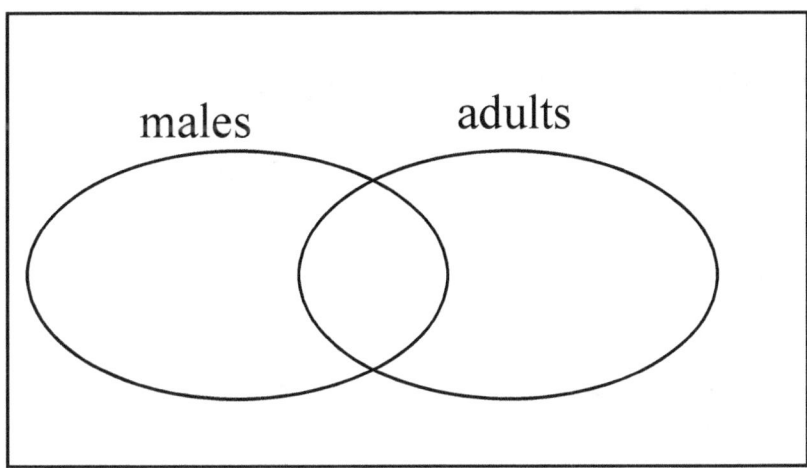

The Venn diagram shows males and adults in a room. There are 10 males in the room who are not adults. There are 5 adult males in the room. Everyone in the room is either a male or an adult. There are totally 25 people in the room. Based on this information, answer the questions listed below.

1) How many adults in the room are not male?

2) How many adults are there in the room?

3) How many males are there in the room?

VENN DIAGRAM

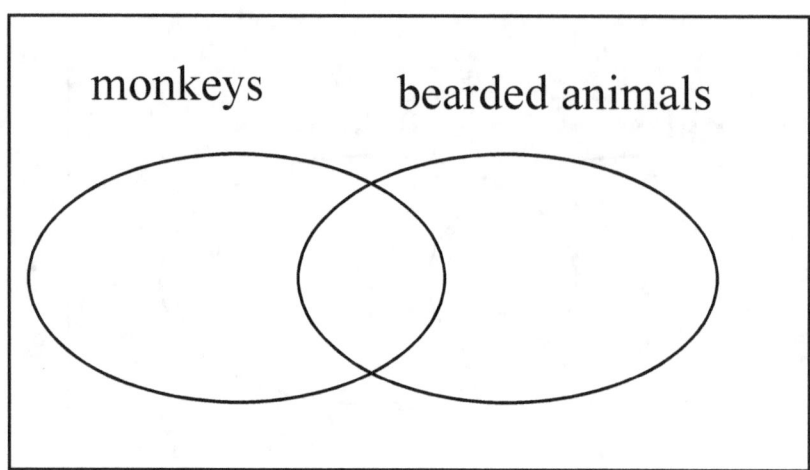

Of the 20 animals in a Zoo, 8 are monkeys. There are 9 animals with beards. 2 animals are neither monkeys nor do they have beards. Based on this information, answer the questions listed below.

1) How many monkeys have beards?

2) How many monkeys do not have beards?

3) How many animals with beards are not monkeys?

4) How many animals are either monkeys or bearded?

Name _____   Date _____

3
VENN DIAGRAM

Group A: Monkeys
Group B: Animals
Group C: Animals that live in water.

Which of the following Venn diagrams represents the relationship between the groups A, B and C listed above?

A)

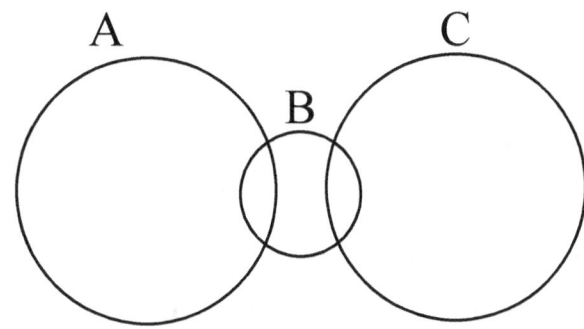

B)

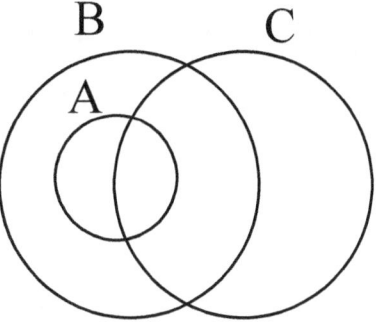

C)

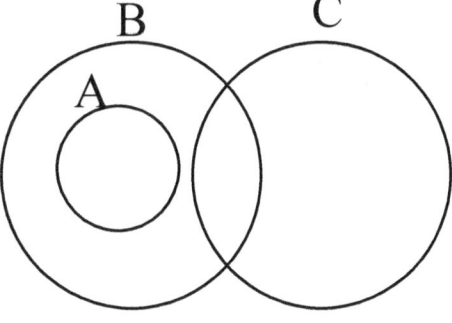

GRAPH LOGIC

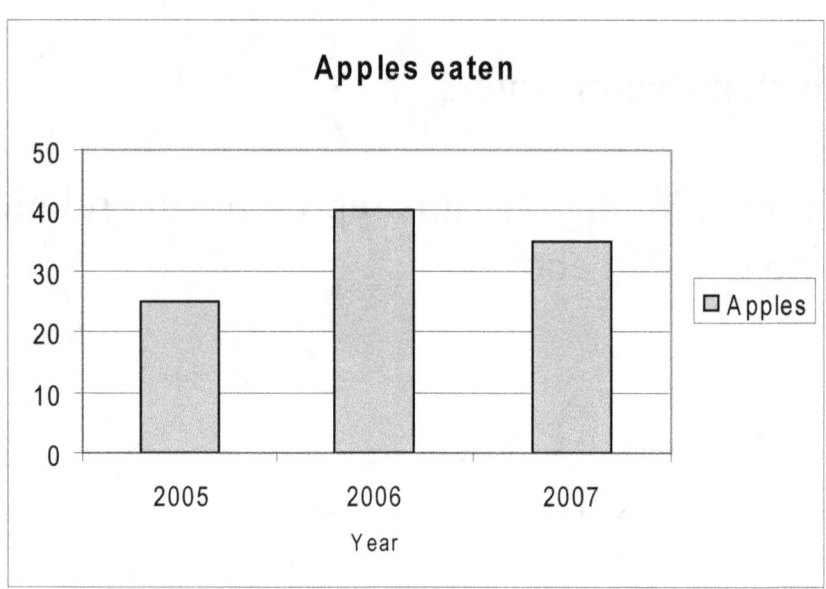

The bar graph above shows the number of apples that Patrick ate during three years. Read the graph and answer the questions.

1) Patrick ate more apples each year compared to its previous year.
 A) True   B) False

2) If Patrick had eaten six more apples in 2007, then 2007 would be the year when he ate the most number of apples.
 A) True   B) False

Name —————————————— Date ——————————

2
GRAPH LOGIC

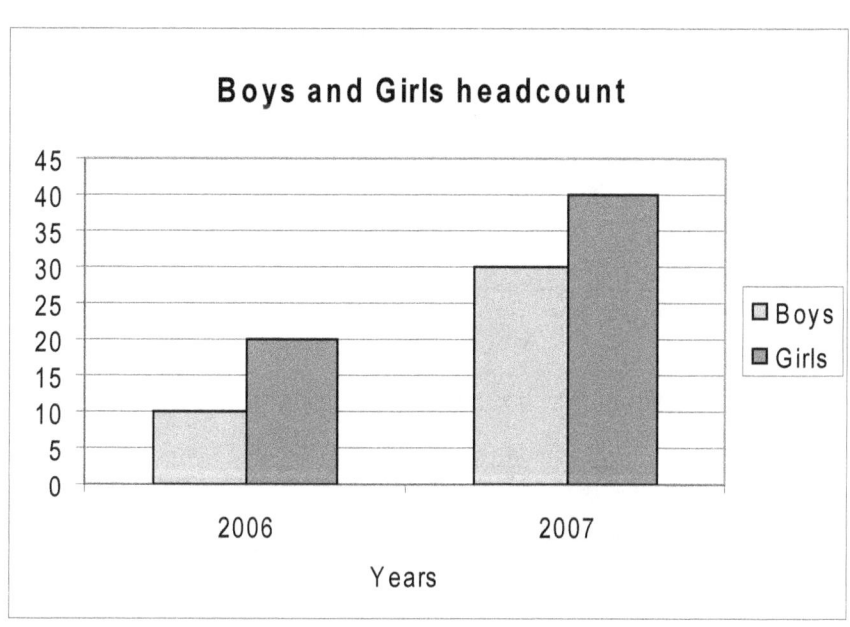

The clustered bar graph above shows the head count of the number of boys and girls in a day care facility. Read the graph and answer the questions.

1) There were more boys in 2007 than in 2006.
    A) True   B) False

2) There were less number of girls in 2007 than in 2006.
    A) True   B) False

Name _____ Date _____

3 GRAPH LOGIC

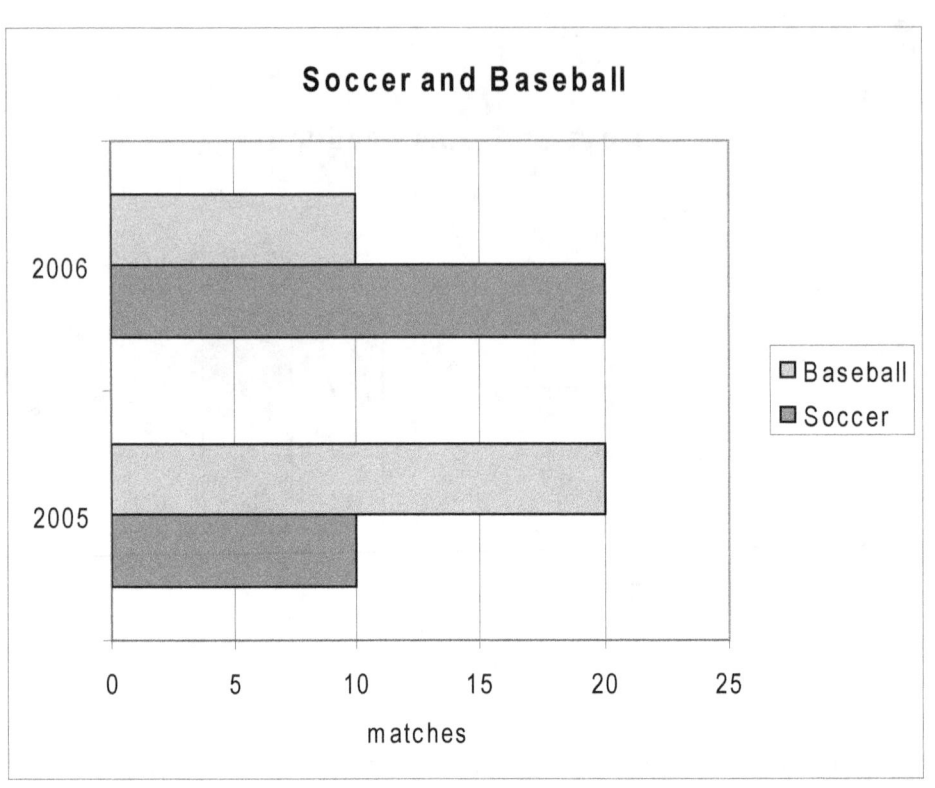

The clustered bar graph shows the number of soccer and baseball games played in a park in two different years. Read the graph and answer the questions.

1) More soccer games were played in 2006 than in 2005.
   A) True   B) False

2) Less number of baseball games were played in 2005 than in 2006.
   A) True   B) False

3) More games were played at the park in 2005 than in 2006.
   A) True   B) False

Name _____   Date _____

4 GRAPH LOGIC

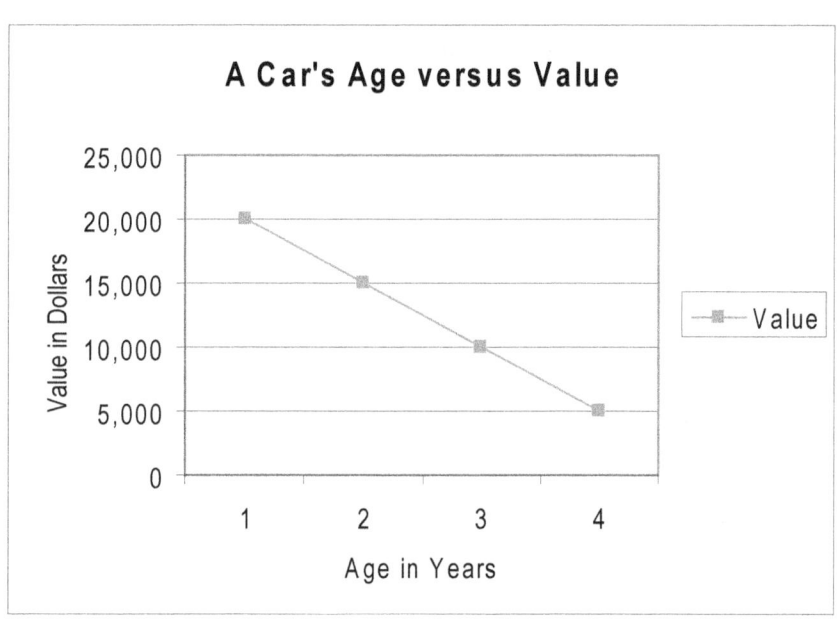

The line graph above shows the value of a car as it ages. Read the graph and answer the questions.

1) As a car becomes older, its value decreases.
   A) True   B) False

2) Newer a car, the more expensive it is.
   A) True   B) False

3) The slump in value from 15,000 to 10,000 dollars started when the car was three years old.
   A) True   B) False

GRAPH LOGIC

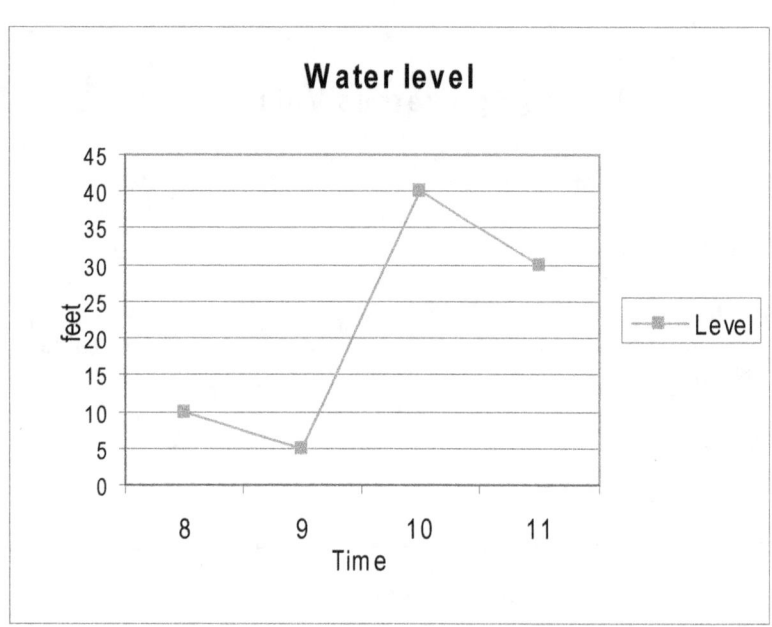

The line graph shows the water level in a tank from 8 AM to 11 AM. Read the graph and answer the questions.

1) Water level was constant between 8 AM and 9 AM.
   A) True   B) False

2) Sometime between 9 AM and 10 AM, the water level in the tank was 41 feet.
   A) True   B) False

3) The water level in the tank dropped by the same amount from 8 AM to 9 AM and from 10 AM to 11 AM.
   A) True   B) False

Name _____   Date _____

6 GRAPH LOGIC

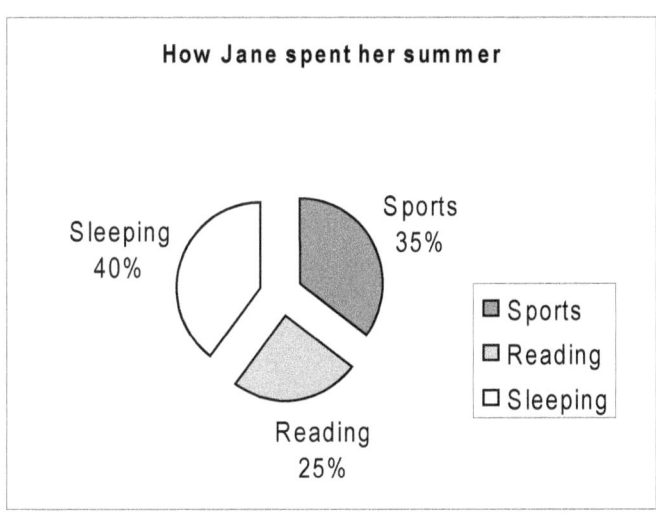

The pie-graph shows the amount of time spent by Jane in three different activities. Read the graph and answer the questions.

1) The predominant activity during the summer was
   A) sports   B) sleeping   C) reading

2) Compared to sports, less time was spent on reading.
   A) True   B) False

3) Jane slept more time than playing sports and reading books combined.
   A) True   B) False

GRAPH LOGIC

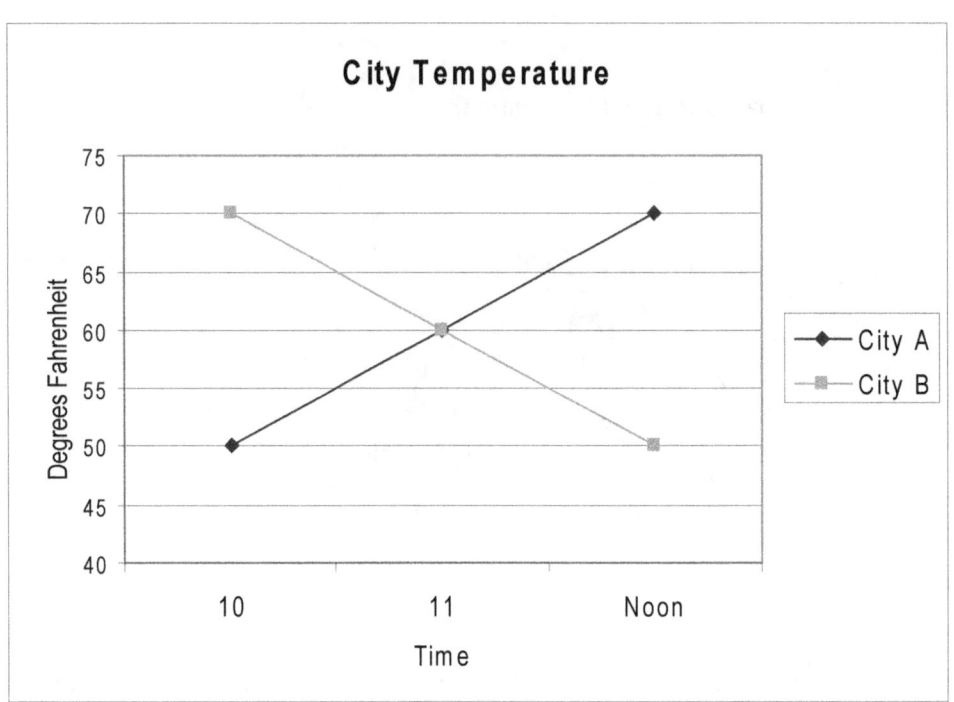

The temperature from 10 AM to Noon in two cities A and B are shown in the line graph. Read the graph and answer the questions.

1) Between 10 and 11 in the morning, city B is cooler than city A.
    A) True   B) False

2) City A is always cooler than city B.
    A) True   B) False

GRAPH LOGIC

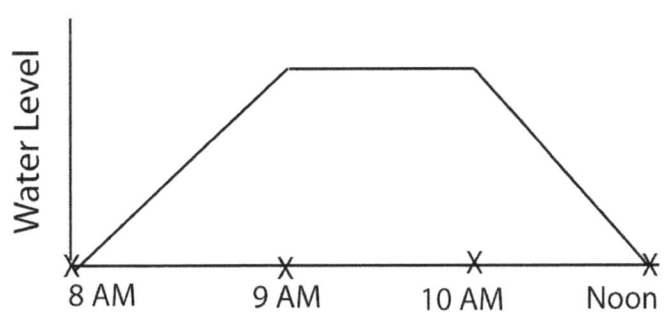

The graph above shows the level of water in a tank. Water is not available from 8 AM to 9 AM when the tank is filled. After 9 AM, the water in the tank can be used. Read the graph and answer the questions.

1) Water was not used from the tank from 9 AM to 10 AM.
   A) True   B) False

2) Water was drained from the tank only from 10 AM to Noon.
   A) True   B) False

NUMBER LOGIC

Figure out the logic in the sequence and find the missing number.

1 ½   1   1½   ?

2 2½   2   1½   ?

3 ¼   ½   ¾   ?

4 4   8   16   ?

5 3   6   ?   24

6 2   1   ½   ?

7 10   5   ?   1¼

8 4   1   ?   1/16

NUMBER LOGIC

Figure out the logic in the numbers and find the missing number.

9

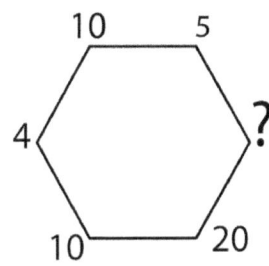

10

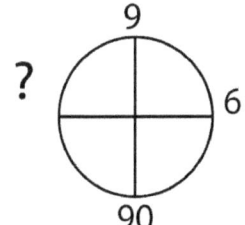

11

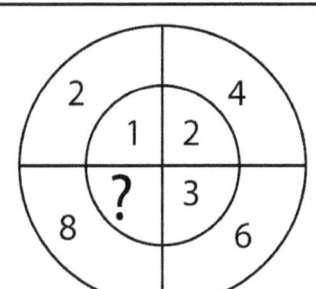

12

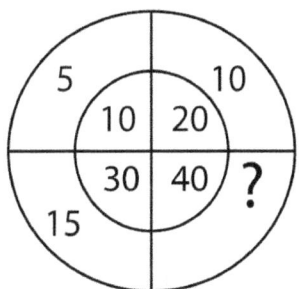

13

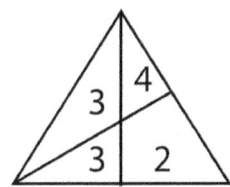

NUMBER LOGIC

Figure out the logic in the numbers and find the missing number.

14

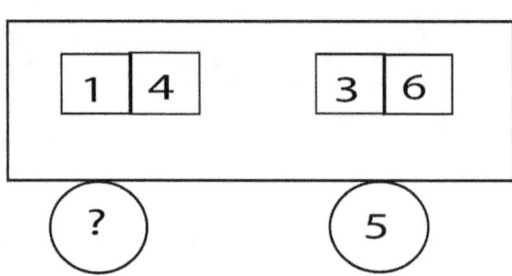

15

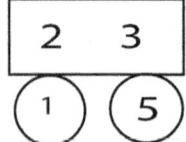 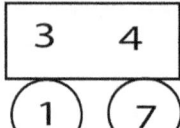 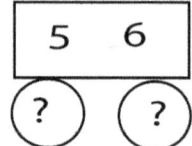

16

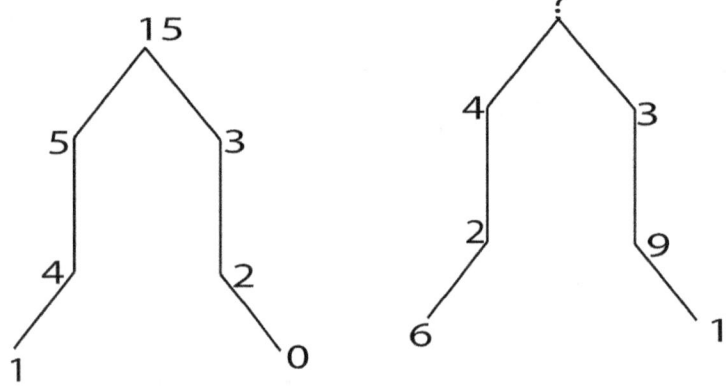

17

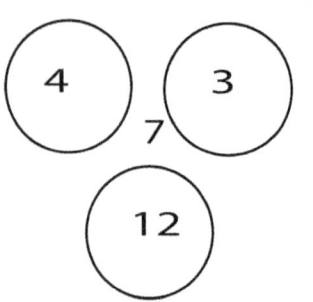 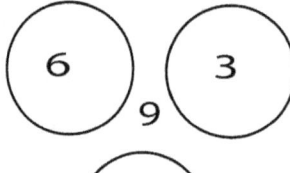

18

100,000   10,000   ?   100

LETTER LOGIC

Figure out the logic in the sequence and find the missing letter or number.

1   A2   B4   C?   D8

2   ACE   MOQ   V??

3   AZ   BY   C?

4   C1E   K2M   Q?S

5   AUG   OCT   ?

6   J   F   M   A   ?

7   SUN   TUE   THU   ?

LETTER LOGIC

Figure out the logic in the sequence and find the missing letter or number.

8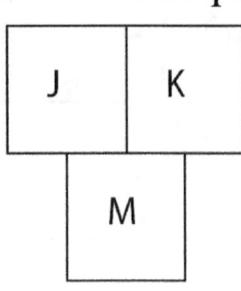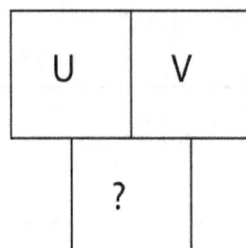

9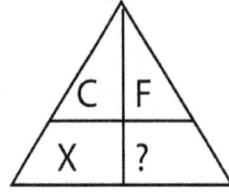

10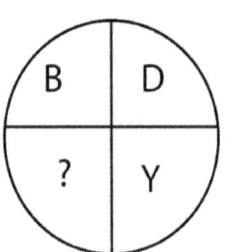

11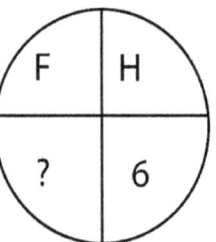

12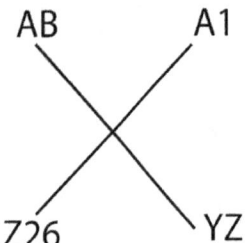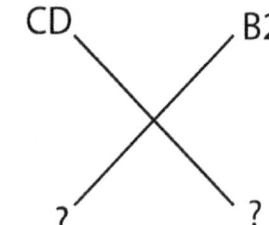

LETTER LOGIC

Figure out the logic in the sequence and find the missing letter or number.

13 MNP   QRT   ?

14 LmN   OpQ   ?

15 A4C   E16K   I?Q

16 A1B2   X24?26

17 Az   By   ?

18 aAz   bBy   ?

19 A^B B^C ?

SUDOKU

Solve the following Sudoku. A correctly solved Sudoku has numbers 1-9 appearing only once in each row, each column and each 3x3 grid. You gain valuable positioning skills by solving these sudokus.

4	1	8	2	9	6	5	3	7
	5	6		8	4	9		1
9	2		3	5	1	6	8	4
2	3	4	5	6	9	7		8
6	7	1	4	3	8	2	5	9
8	9	5	1		7	3	4	6
7		2	9	4	5	1	6	3
5	4	9	6	1	3	8		2
1	6			7		4	9	5

Analytical Reasoning
© Gift Of Logic, Inc * Copying prohibited

SUDOKU

Solve the following Sudoku. A correctly solved Sudoku has numbers 1-9 appearing only once in each row, each column and each 3x3 grid. You gain valuable positioning skills by solving these sudokus.

2		3	5		8	1		4
5	8		4	9		2	6	
6	4	1		3	2	5		8
8	5	4	6	1	9	7	3	2
	7	2	8	4			1	9
1	6	9	2	7	3	4	8	5
9	2		1	8	7	3	4	6
4	1	8	3		6	9		7
7	3	6	9	2	4	8	5	1

Analytical Reasoning

SUDOKU

Solve the following Sudoku. A correctly solved Sudoku has numbers 1-9 appearing only once in each row, each column and each 3x3 grid. You gain valuable positioning skills by solving these sudokus.

4	7	9	1		8	6	3	5
	2	3	9	4	6	8	7	1
8	6	1	5	7	3	2	9	
3	8	4	2	5	1	9	6	7
9	5	2		6	4	3		8
6		7	3	8	9	5	4	2
2	4	5	6	9	7		8	3
1	9	8		3	2	7	5	6
7		6	8	1	5	4	2	9

Name _____   Date _____

4
SUDOKU

Solve the following Sudoku. A correctly solved Sudoku has numbers 1-9 appearing only once in each row, each column and each 3x3 grid. You gain valuable positioning skills by solving these sudokus.

8		2	7	5	3	4	6	1
1	7	6	2	4	9		8	5
3	5	4	8		6	9	7	2
4	6		5	9	2	7	1	3
9	2	5	3	7		6		8
7		1	6		4	2	5	9
5	4	7	9	3	8	1	2	6
2		9	4	6		8	3	7
6	8	3	1	2	7		9	4

Analytical Reasoning
© Gift Of Logic, Inc * Copying prohibited

SUDOKU

5

Solve the following Sudoku. A correctly solved Sudoku has numbers 1-9 appearing only once in each row, each column and each 3x3 grid. You gain valuable positioning skills by solving these sudokus.

8		7	1	6	4	2	3	9
9	1	6	2	3	8	7		4
3	2		9		7		8	6
	8	9	7	1	2	3		5
5	7	1			6	4	9	2
2		3	5	4	9		1	7
	9	8	4	7	5	6	2	3
7	3		6	9		5		8
6	4	5	8	2	3	9	7	1

Analytical Reasoning

6 SUDOKU

Solve the following Sudoku. A correctly solved Sudoku has numbers 1-9 appearing only once in each row, each column and each 3x3 grid. You gain valuable positioning skills by solving these sudokus.

3		8	1	9	5	7	2	6
5	2		4		6	1		3
1	7	6	2	3	8	5	4	9
	9	1	8	2			5	4
6	8	3			7	2		1
4	5		9	6	1	3	7	8
9		4	3	5	2		1	7
2	1	7	6		9	4		5
8	3	5	7	1	4	9	6	2

Analytical Reasoning

Name ——————————— Date ———————————

PICTORIAL REASONING

Name _____   Date _____

PICTURE SEQUENCE

Figure out the logic in the picture sequence, and draw the next picture in the sequence.

1

2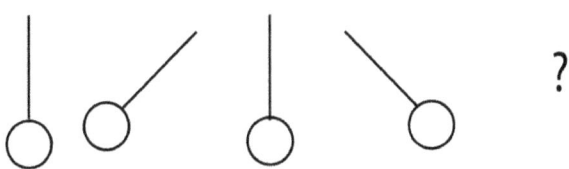

3

4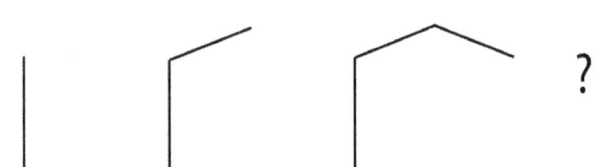

5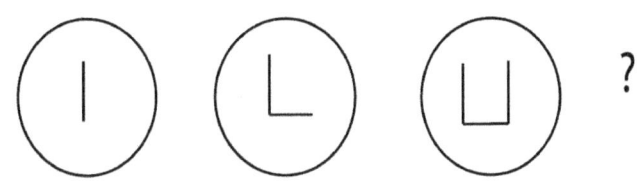

6

Pictorial Reasoning Answers-137
© Gift Of Logic, Inc * Copying prohibited

Name _____   Date _____

PICTURE SEQUENCE

Figure out the logic in the picture sequence, and draw the next picture in the sequence.

7 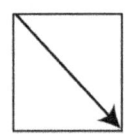 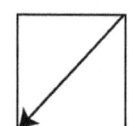 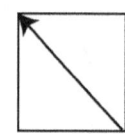 ?

8 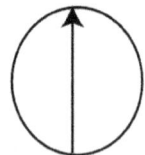 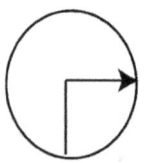 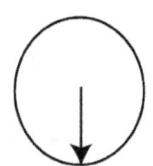 ?

9 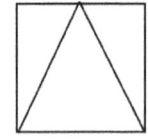 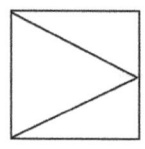 ?

10  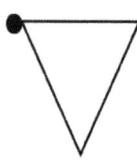 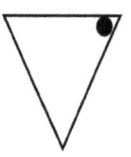 ?

11 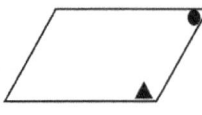 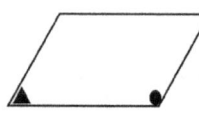  ?

Pictorial Reasoning   Answers-138   74
© Gift Of Logic, Inc * Copying prohibited

Name _____   Date _____

PICTURE SEQUENCE

Figure out the logic in the picture sequence, and draw the next picture in the sequence.

12 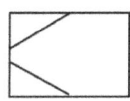 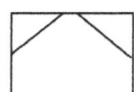  ?

13 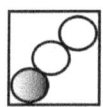 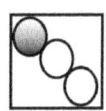 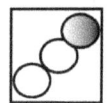 ?

14 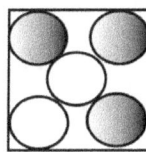 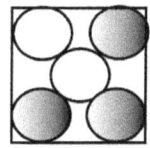 ?

15    ?

16 ↑→ →↓ ←↓ ?

Pictorial Reasoning   Answers-139
© Gift Of Logic, Inc * Copying prohibited

Name _____   Date _____

PICTURE ANALOGY

Figure out the logic in the picture analogy, and circle the correct picture that will complete the analogy.

1 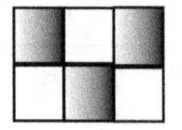 AS 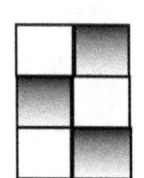 :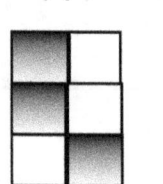

2 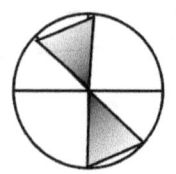 AS 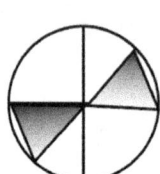 :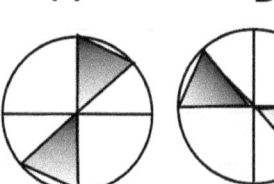

3  AS  :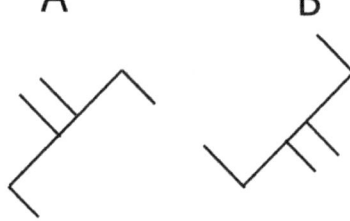

4  : ... AS ... : A B C

Pictorial Reasoning   Answers-140   76
© Gift Of Logic, Inc * Copying prohibited

PICTURE ANALOGY

Figure out the logic in the picture analogy, and circle the correct picture that will complete the analogy.

5

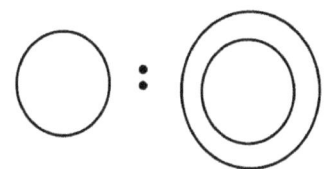 AS 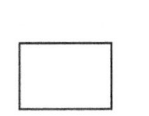 :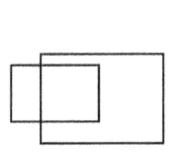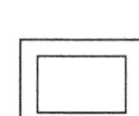

A B C

6

 AS 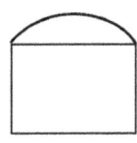 :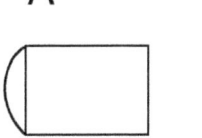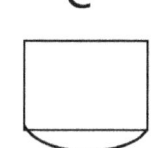

A B C

7

 AS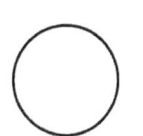

A B

8

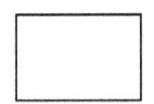 AS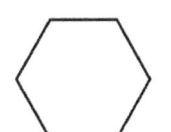

A B

Pictorial Reasoning   Answers-141

PICTURE ANALOGY

Figure out the logic in the picture analogy, and circle the correct picture that will complete the analogy.

9   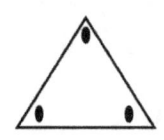 :  AS 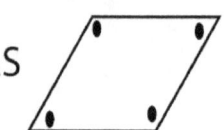 :   A   B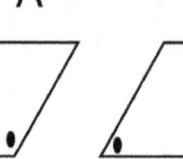

10   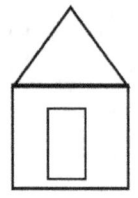 AS 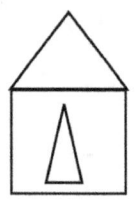 :   A   B   C 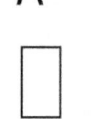 

11   AS  :   A   B   C

12    :  AS 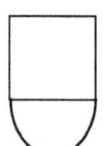 :   A   B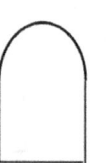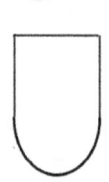

13   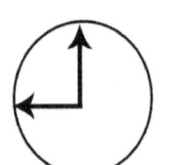 : 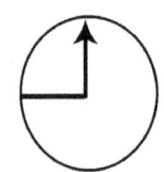 AS 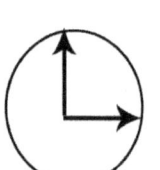 :   A   B   C

Pictorial Reasoning   Answers-142   78

Name _____ Date_____

ODD PICTURE

Figure out the logic in the pictures and identify the odd picture.

1   A   B   C

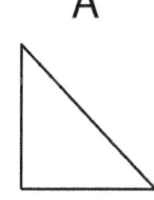  

2   A   B   C

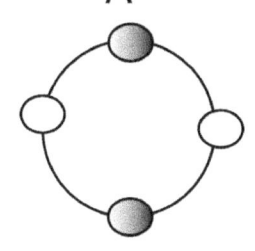 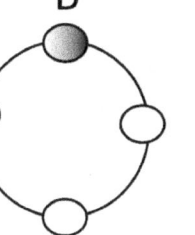 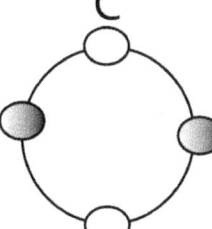

3   A   B   C

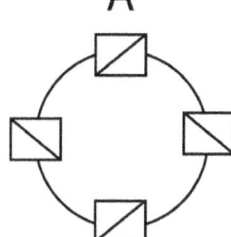 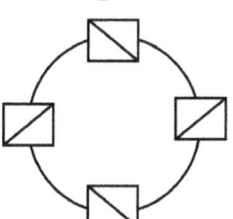 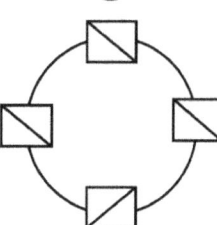

4   A   B   C

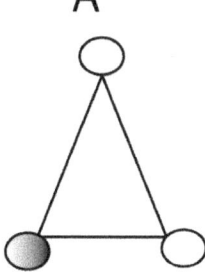 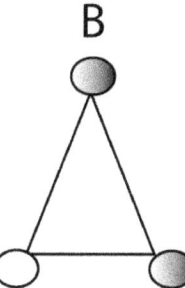 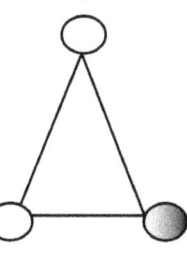

5   A   B   C

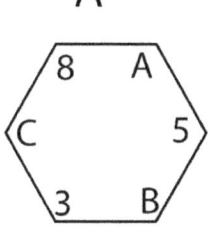 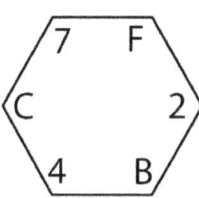

Pictorial Reasoning   Answers-143
© Gift Of Logic, Inc * Copying prohibited

Name _____   Date _____

ODD PICTURE

Figure out the logic in the pictures and identify the odd picture.

6 A B C
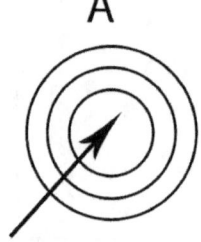 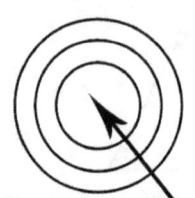 

7 A B C
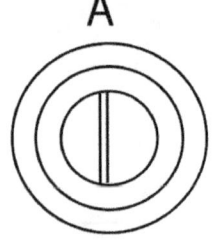 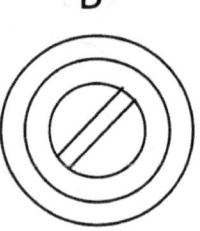 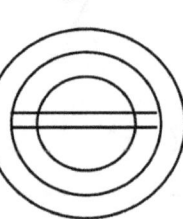

8 A B C
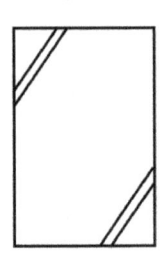  

9 A B C
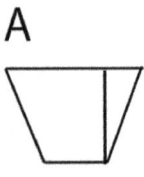 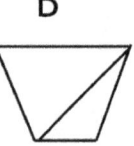 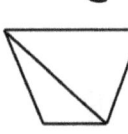

10 A B C

Pictorial Reasoning   Answers-144
© Gift Of Logic, Inc * Copying prohibited

Name _____   Date _____

PICTURE DIFFERENCE

Mark the differences in the set of pictures shown, with arrows.

1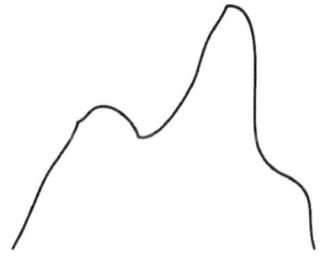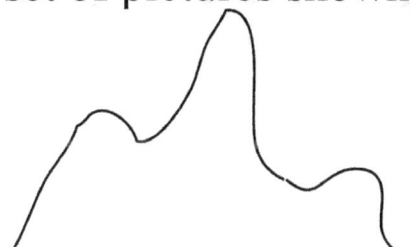

2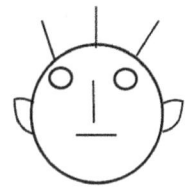

3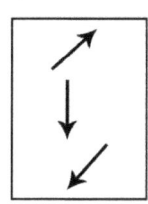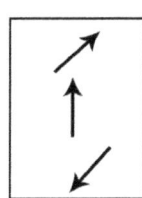

4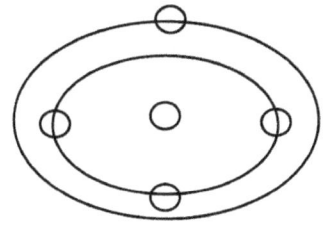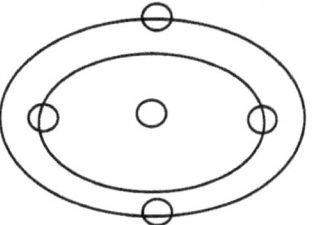

5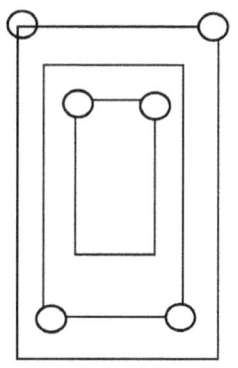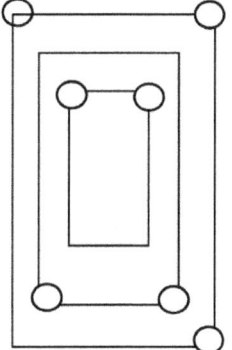

Name _____   Date _____

PICTURE DIFFERENCE

Mark the differences in the set of pictures shown, with arrows.

6

7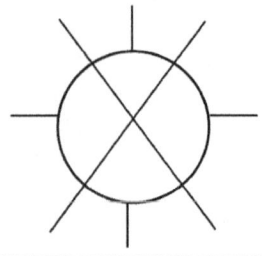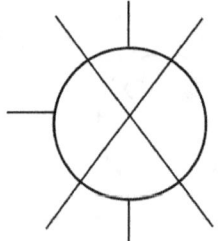

8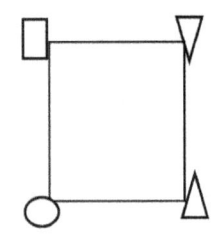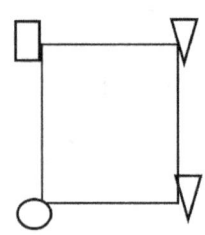

9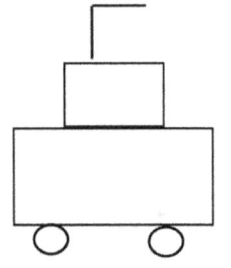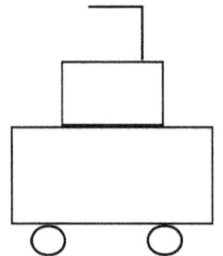

10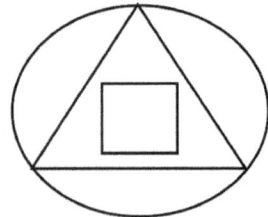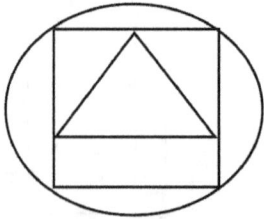

Pictorial Reasoning   Answers-146
© Gift Of Logic, Inc * Copying prohibited

Name _____ Date _____

PICTURE DIFFERENCE

Mark the differences in the set of pictures shown, with arrows.

11

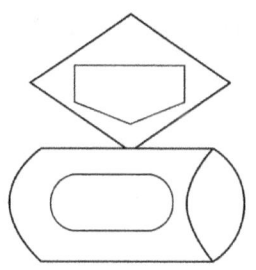 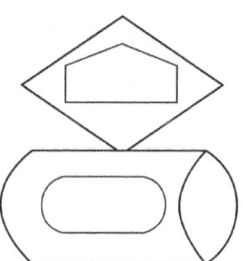

12

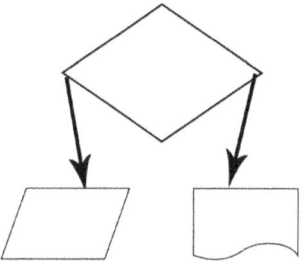 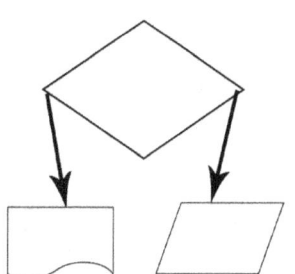

13

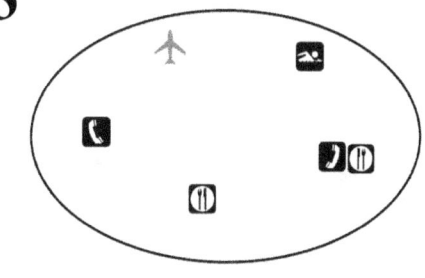 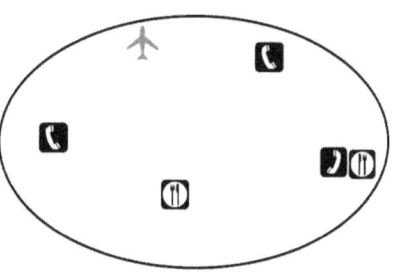

14

15

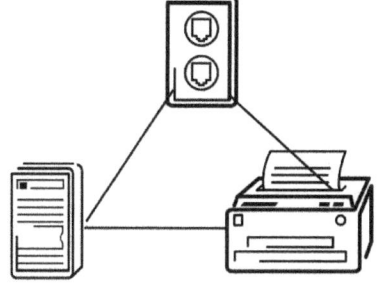 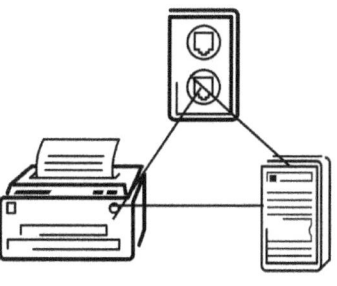

Pictorial Reasoning   Answers-147
© Gift Of Logic, Inc * Copying prohibited

Name _____   Date _____

PATTERN MATCHING

Find the logical pattern in the pictures on the left, and identify the picture on the right that will fit in the space marked with ? to complete the pattern.

1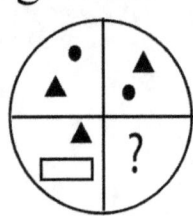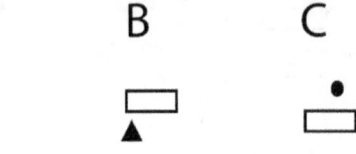

2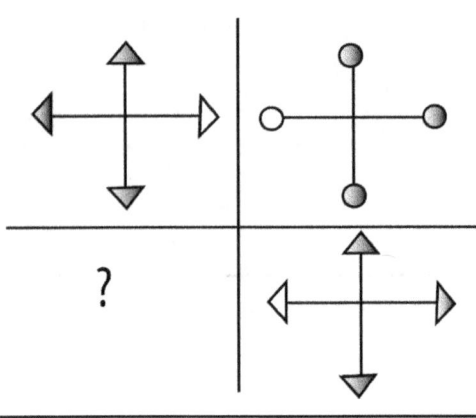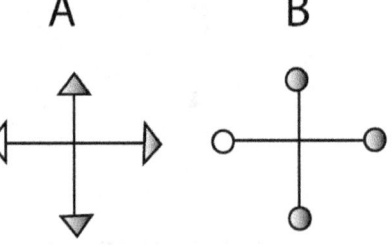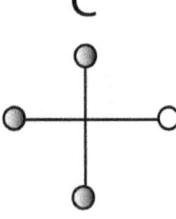

3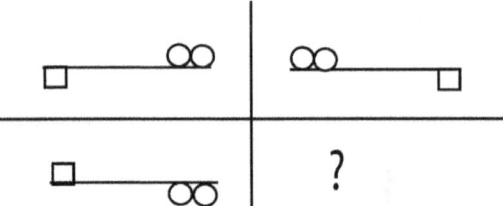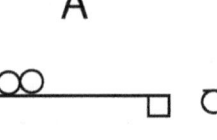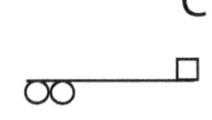

4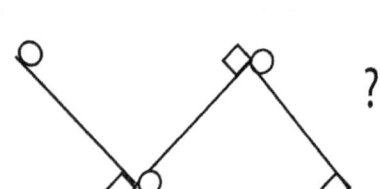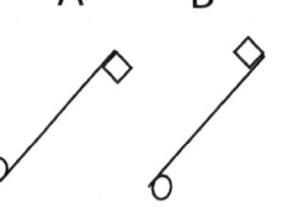

Pictorial Reasoning
© Gift Of Logic, Inc * Copying prohibited

ANSWERS

CONVERSE STATEMENTS - conditional

Answer the following questions and explain your reasoning.

1 Conditional: If a movie is good, people will watch it.
   Inference: If people watch a movie, then it must be a good movie.
   Is the inference valid?   Answer: B) No

Reasoning: The inference is the converse of the conditional. The converse of a conditional is a false inference.
   conditional: movie is good → people will watch it
   converse: people watch a movie → movie is good.
Just because people watch a movie does not mean that it will be good.

---

2 Conditional: If it is 10 PM, the theater will close.
   Converse: If the theatre closes, then it is 10 PM.
Is the converse a valid inference? Answer: B) No
Reasoning: The converse of a conditional is an invalid inference. The theater could close at anytime - say for an emergency at 9 PM.

---

3 Conditional: snake bites → will hurt
   Converse: hurts → snake bite
Is the converse a valid inference? Answer: B) No
Reasoning: You can get hurt by falling down, but that does not mean that a snake bit you!

---

4 Conditional: very warm → plants will die
   Converse: plants will die → very warm
Is the converse a valid inference? B) No
Reasoning: Plants could die because of various reasons other than warm temperatures. For example, they could die because of very cold weather.

Answers
© Gift Of Logic, Inc * Copying prohibited

CONVERSE STATEMENTS - unconditional

Write the converse of the given statements and find whether it is true or false.

1   Statement: All books are study materials.
       Converse: All study materials are books.
The converse is a B) Invalid inference.
<u>Reasoning:</u> Some study materials can be CDs, charts etc.

2   Statement: The red string is tied to the blue string.
       Converse: The blue string is tied to the red string
The converse is a A) Valid inference
<u>Reasoning:</u> Since both strings are "tied" together, the converse must be true as well.

3   Statement: The boat race was held after the swimming race.
       Converse: The swimming race was held after the boat race.
The converse is a B) Invalid inference
<u>Reasoning:</u> The timing of the races are important. The boat race was held "after" the swimming race, not the other way around.

4   Statement: Abdul met Ali.
       Converse: Ali met Abdul.
The converse is a A) Valid inference. <u>Reasoning:</u> When A meets B, it is obvious that B also meets A.

5   Statement: Kathy and Cindy are friends. Converse: Cindy and Kathy are friends. The converse is a A) Valid inference <u>Reasoning:</u> When A and B are friends, it is correct to infer that B and A are friends as well.

Answers

INVERSE STATEMENTS

Conditional: P → Q
Inverse: ~P → ~Q
Inverse of a conditional is an invalid inference

---

1 Conditional: If the building is old, it will be painted.
   Inverse: If the building is not old, it will not be painted.
   Conditional Symbolic: building is old → painted
   Inverse Symbolic::   ~ building is old → ~ painted

Is the inverse true or false? B) False.

Reasoning: A new building could be painted. The condition is regarding old buildings only.

2

Statement 1: If the pencil is sharp, the handwriting will be good.
Statement 2: The pencil is not sharp.
Statement 3: The handwriting will not be good.

Does statement 1 and statement 2 together imply statement 3?
   B) No

Reasoning: Even if the pencil is not sharp, the handwriting can be good if extra care is taken. The condition means that if the pencil is sharp, then the handwriting will be good. We cannot infer that, if the pencil is not sharp, then the handwriting will not be good.

Answers
© Gift Of Logic, Inc * Copying prohibited

INVERSE STATEMENTS

3
    Conditional: The alarm will sound if there is a fire.
    Inverse: If there is no fire, the alarm will not sound.
    Conditional Symbolic: fire → alarm
    Inverse Symbolic: ~ fire → ~alarm

Is the inverse true or false? B) False

Reasoning: Inverse of a conditional is a false inference. The alarm could malfunction and sound even if there is no fire.

---

4   Conditional: If the soil is not fertile, the plants will not grow.
    Inverse: If the soil is fertile, the plants will grow.
    Conditional Symbolic: ~ fertile → ~ grow
    Inverse Symbolic: fertile → grow

Is the inverse true or false?   B) False

Reasoning: Note that even though the inverse sounds reasonable, it is a false inference. We only know from the conditional that if the soil is not fertile, it will not grow. We cannot say for sure that if the soil is fertile, then the plants will grow. Even if the soil is fertile, it is possible that the plants do not grow for some reason.

Note that in the answers to questions 1-4, even though we provide detailed reasoning, we can answer whether the inverse is true or false instantly. The inverse of a conditional statement is always a false inference.

Answers
© Gift Of Logic, Inc * Copying prohibited

CONTRAPOSITIVE STATEMENTS

Write the contrapositive statements for the following conditionals.

1   Conditional: If there is inflation, prices will go up.
   Contrapositive: If the prices do not go up, then there is no inflation.
Symbolic:
   Conditional: inflation → prices go up
   Contrapositive: ~prices go up → ~inflation

Is the contrapositive a valid inference? A) Yes

---

2   Conditional: If it is foggy, then the airplane cannot land.
   Contrapositive: If the airplane can land, then it is not foggy.
Symbolic:
   Conditional: foggy → ~plane land
   Contrapositive: plane land → ~foggy

Is the contrapositive a valid inference? A) Yes

3

Conditional: If the weather is good, the rocket will take off.
Contrapositive: If the rocket does not take off, the weather is not good.

Symbolic:
   Conditional: weather is good → rocket will take off
   Contrapositive: ~rocket take off → ~weather is good

Is the contrapositive a valid inference? A) Yes

---

answers

© Gift Of Logic, Inc * Copying prohibited

CONTRADICTORY STATEMENTS - conditional

If then: P → Q
Contradiction: P → ~Q

If a conditional statement is true, then the contradiction is false. It is logically impossible for the contradiction to be true.

---

1 Conditional: If you want dinner, you must come home.
Symbolic: dinner → come home

Contradiction: If you want dinner, you must not come home.
Symbolic: dinner → ~come home

Contradiction is B) False

---

2 Conditional: If you are talented, you will be popular.
Symbolic: talented → popular

Contradiction: If you are talented, you will not be popular.
Symbolic: talented → ~popular

Contradiction is B) False

Answers
© Gift Of Logic, Inc * Copying prohibited

CONTRADICTORY STATEMENTS - unconditional

Statements that do not have conditions also can be contradictory.

A statement is said to be contradictory or inconsistent if it is logically impossible for it to be true.

---

1 Make up a contradictory statement:

I like to be the president of my country, but I don't like to be a public servant.

---

2 Make up a contradictory set of statements:

Some people have climbed Mount Everest.
No one has climbed Mount Everest.

I obey the law at all times.
I drove above the speed limit.

I like fruits.
I hate apples.

Answers

© Gift Of Logic, Inc * Copying prohibited

INFERENCING - must be true

1 If the weather is good, the airplane must land.

A) If the airplane lands, then the weather is good.
B) If the airplane does not land, then the weather is not good. (Answer)

Reasoning:

Condition: weather is good → airplane will land
Converse: airplane lands → weather is good
Contrapositive: ~airplane lands → ~weather is not good

Choice A is the converse of the conditional and therefore cannot be true.
Choice B is the contrapositive of the conditional and therefore, it must be true.

---

2 If a book sells a lot, it must be a good book.

A) If a book sells a lot, it must not be a good book.
B) If a book is a good book, it will sell a lot.
C) If a book does not sell a lot, it must not be a good book.
D) If a book is not a good book, it will not sell a lot. (Answer)

Reasoning:

sells a lot → good book
~good book → ~sell a lot

A - incorrect - this is the contradiction of the conditional - so, it is false.
B - incorrect - this is the converse of the conditional - so, it is false.
C - incorrect - this is the inverse of the conditional - so, it is false.
D - correct - this is the contrapositive of the conditional - so, it is true.

Answers
© Gift Of Logic, Inc * Copying prohibited

INFERENCING - must be true

3 When the principal speaks, everyone must listen.

A) When the principal does not speak, everyone will not listen.
B) If everyone is listening, the principal must be speaking.
C) If everyone is not listening, then the principal must not be speaking. (Answer)

Reasoning: principal speaks → everyone must listen
~everyone is listening → ~principal speaks (contrapositive)

A - incorrect - inverse of the conditional
B - incorrect - converse of the conditional
C - correct - contrapositive of the conditional is a valid inference and so, it must be true.

---

4 If apples are grown in a farm, then oranges must also be grown. This year, oranges were not grown in the farm.

A) apples were grown this year.
B) apples were not grown this year. (Answer)

Reasoning: apples grown → oranges grown (conditional)
~oranges grown → ~apples grown (contrapositive, true)

Since we know that oranges were not grown this year, from the contrapositive, we can infer that apples were not grown.

Answers
© Gift Of Logic, Inc * Copying prohibited

INFERENCING - must be true

5 A car mechanic found two problems with Jack's car. The problems are such that if the first problem is fixed, then the second problem must also be fixed.

    A) The mechanic fixed the second problem only. (Answer)
    B) The mechanic fixed the first problem, but not the second problem.

Reasoning: fix first problem → fix second problem

Choice B cannot be correct since it is a contradiction to the given condition. Choice A is correct. Remember that "If P then Q" does not imply "If Q then P". So, it is possible that only the second problem got fixed.

---

6 Amanda must go to the wedding if Bobby also goes. But, Amanda could not attend the wedding due to an emergency.

    A) Bobby attended the wedding.
    B) Bobby did not attend the wedding. (Answer)

Reasoning: Bobby goes to wedding → Amanda goes to wedding
               ~Amanda goes → ~Bobby goes (contrapositive)

Choice B is the correct answer as it is the contrapositive of the condition. Choice A contradicts the contrapositive statement and so, it is incorrect.

INFERENCING - must be true

7 If the bus is on time, Jane must not take the train to work. But, on Monday, Jane took the train to work.

   A) The bus was on time
   B) The bus was not on time. (Answer)

<u>Reasoning:</u>   bus on time → ~take train
                    take train → ~bus on time

Jane took the train to work. From the contrapositive, it must be true that the bus was not on time. Choice B is the correct answer.

---

8 A party was scheduled to begin at 6 PM to honor Jack and Kate for their outstanding achievements. The party must not begin unless Jack and Kate arrives. The party began at 7 PM instead of 6 PM.

   A) Either Kate or Jack or both did not arrive before 7 PM.
   B) Both Kate and Jack arrived before 7 PM. (Answer)

<u>Reasoning:</u> " The party must not begin unless Jack and Kate arrives" can be rewritten using "if-then" as follows:
   If Jack and Kate do not arrive then the party must not begin.
   ~Jack and Kate arrive → ~party begin
   party begins → Jack and Kate arrived (contrapositive)

We know that the party began at 7 PM instead of 6 PM. So, it must be true that both Jack and Kate arrived before 7 PM.

Answers
© Gift Of Logic, Inc * Copying prohibited

INFERENCING - must be true

9 If Arthur wants to buy a camera, he must buy it from shop A or shop B only. But, Arthur was not able to buy a camera.

A) Shop A had the camera that Arthur wanted, but shop B did not.
B) Shop A and shop B both did not have the camera that Arthur wanted. (Answer)

Reasoning: buy camera → shop A or shop B
 ~ (shop A or shop B) → ~buy camera

Choice A is incorrect since if shop A had the camera, Arthur would have bought it. Choice B is correct - since both the shops did not carry it, he was not able to buy it. Note that the contrapositive of the condition helps us to verify that choice B is the correct answer.

INFERENCING - cannot be true

10 If Sue is sick, she must not delay her visit to the doctor.

A) If Sue is not delaying her doctor's visit, she must be sick. (Answer)
B) If Sue is delaying her doctor's visit, she must not be sick.

Reasoning: conditional: sick → ~delay doctor's visit
 converse: ~delay doctor's visit → sick
 contrapositive: delay doctor's visit → ~sick

Choice A - correct answer. This is the converse of the conditional, which cannot be true. The choice that cannot be true is the correct answer.

INFERENCING - cannot be true

11 If you wear a tie, you must be a boy.

  A) Nick wore a tie, so he must be a boy.
  B) Nick is a boy, so he must be wearing a tie. (Answer)

Reasoning: wear tie → boy
                boy → wear tie (converse, invalid)

Choice B is the converse of the conditional, which cannot be true. It is ok for Nick to be a boy, but not wear a tie.

---

12 If George wants to play well, he must wear a shoe with cleats.

  A) If George is not playing well, he is not wearing a shoe with cleats. (Answer)
  B) If George is not wearing a shoe with cleats, he is not playing well.

Reasoning: play well → cleated shoes
                ~cleated shoes → ~playing well (contrapositive, valid)
                ~play well → ~cleated shoes (inverse, invalid)

Choice A is the inverse of the conditional and this is an invalid inference. So, this choice cannot be true.

Choice B is the contrapositive and therefore, it must be true.

answers
© Gift Of Logic, Inc * Copying prohibited

INFERENCING - cannot be true

13 If the food is not fresh, you must not eat it.

    A) Food that is not fresh must be eaten. (Answer)
    B) If you are eating the food, it must be fresh.

Reasoning: ~fresh food → ~eat
                eat → fresh food (contrapositive, valid)
                ~fresh food → eat (contradiction, invalid)

Choice A is a contradiction of the conditional and hence cannot be true. So, this is the correct answer. Choice B is the contrapositive of the conditional, which is true.

14 You cannot be punctual unless you wear a watch.

    A) If you are punctual, you must be wearing a watch.
    B) Even without wearing a watch you can be punctual. (Answer)

Reasoning: If you do not wear a watch, you cannot be punctual
               ~wear a watch → ~punctual
               ~wear a watch → punctual (contradiction, invalid)
               punctual → wear a watch (contrapositive, valid)

Choice B is a contradiction of the conditional and cannot be true. So, this is the correct answer. Choice A is the contrapositive and so it must be true.

Answers

AGREE

1 We must balance work and play. Too much work or too much play is not going to do us any good.

Agree: Yes, balancing work and play is very important. Without this balance, we will not be well rounded.

---

2 This lake is very deep. So, it is better that nobody swims in this lake without a lifeguard being present.

Agree: Lifeguards are trained to pull people out of danger. So, it is better not to swim in this deep lake without a lifeguard.

---

3 Water is essential for living. So, sprinkling water in our lawns to keep them green must not be more important than using it for drinking.

Agree: Maintaining a pretty lawn is nice, but water is becoming scarce and must be used for drinking first before using it for other purposes.

---

4 People who litter the streets must be forced to pay a heavy fine.

Agree: Littering makes the streets dirty and it costs money to clean up. So, whoever litters must pay for the cost of its cleanup.

Answers
© Gift Of Logic, Inc * Copying prohibited

DISAGREE

1 It is okay for visitors to feed zoo animals.
Disagree: Zoo animals must only be fed by Zoo authorities.

2 All students who are in the same class are of the same height.

Disagree: Students in the same class are rarely of the same height.

3 In order to live on this earth, everyone must speak two languages.
Disagree: Speaking in one language is sufficient to live on this earth. Most people in the world speak only one language.

4 Cars are fun to drive. So, we should get rid of motorbikes.

Disagree: We cannot get rid of motorbikes because they pollute less than cars and are preferred by those who do not like to drive cars or afford one.

5 Even if a movie is bad, we must watch it anyway if others watch it.

Disagree: If we think that a movie is bad, we should not watch it regardless of what others do.

6 This road is scenic with mountains along the way. So, this road will take us to our destination in the shortest amount of time.

Disagree: Just because it is scenic does not mean that it will be the shortest route to our destination.

Answers

SENTENCE ANALOGY

1 All the lies that Joe has been telling fell apart like a tower of cards.

What are being compared in the statement?
   Lies that Joe has been telling and a tower of cards.

Rewrite the statement without using analogy.
   All the lies that Joe has been telling were discovered quickly.

2 The cheeks of the baby feel like cushion.

What are being compared in the statement?
      Cheeks of the baby and cushion.

The cheeks of the baby are rough. Answer: B) False Since cushion is soft, the cheeks of the baby are also soft.

3 Daniel lived like a frog in a well.

What are being compared in the statement?
      Daniel's life and a frog's life.
Daniel travelled to many places. Answer: B) False Since frogs in a well do not travel to other places, Daniel also did not travel to many places.

4 Victor ran like a tiger.
What are being compared in the statement?
      Victor's speed of running and Tiger's speed of running.
Victor ran Answer: A) very fast
      Since tigers run very fast, Vinod also ran very fast.

Answers

WORD ANALOGY

Q#	Answer	Reasoning
1	B	Pen is to write as Eraser is to clean.
2	B	Toyota is to car as Boeing is to airplane.
3	B	Zoo is to animals as Aquarium is to fish.
4	C	Airport is to planes as Shipyard is to ships.
5	C	USA is to country as Europe is to continent.
6	A	Earth is to Sun as Moon is to Earth. Earth goes around the Sun. Moon goes around the earth.
7	B	Palace is to king as Jail is to convict.
8	B	Lion is to forest as Fish is to water.
9	C	Air is to breathe as Water is to drink.
10	C	Car : drive :: Bike : ride
11	C	Piano : music :: Paint : drawing
12	B	State : governor :: City : mayor
13	A	Dog : bark :: Cat : meow
14	C	Gold : costly :: Dirt : cheap
15	B	North : south :: East : west
16	B	Apple : fruit :: Carrot : vegetable
17	B	Nurse : doctor :: Mechanic : engineer
18	A	Body : skeleton :: Building : frame

1 LIST PROCESSING - sorting based on one property

The names of five people..

Name	Height	Weight
Andy	4	30
Bobbie	3	50
Cathy	4	40
Dinesh	3	30
Emma	5	20

Sorting and ranking based on height in descending order. Note that Andy and Cathy have the same rank because their height is the same.

Name	Rank
Emma	1
Andy	2
Cathy	2
Bobbie	3
Dinesh	3

Sorting and ranking based on weight in descending order. Note that Andy and Dinesh have the same rank because their weight is the same.

Name	Rank
Bobbie	1
Cathy	2
Andy	3
Dinesh	3
Emma	4

Why are the ranks different in the two lists? One is sorted based on height and one is sorted based on weight.

answers
© Gift Of Logic, Inc * Copying prohibited

2   LIST PROCESSING - sorting based on two properties

Now, sort the names (from the first table) again, first in descending order of height and if there is a tie, sort on weight in descending order to resolve the tie.

Name	Rank
Emma	1
Cathy	2
Andy	3
Bobbie	4
Dinesh	5

Since Andy and Cathy have the same height, there is tie. So, we break the tie by using their weight - Cathy is 40 lbs and Andy is 30 lbs. So, Cathy will appear before Andy in the list. The tie between Bobbie and Dinesh is also broken using their weight - Bobby is heavier than Dinesh.

Sort the names (from the first table), first in descending order of weight and if there is a tie, sort on height in descending order to resolve the tie.

Name	Rank
Bobbie	1
Cathy	2
Andy	3
Dinesh	4
Emma	5

Since Andy and Dinesh have the same weight, we break the tie using their heights - Andy is 4 ft, but Dinesh is 3 ft. So, Andy appears before Dinesh in the list.

Answers

3 LIST PROCESSING - adding to a list

A list is sorted in descending order and currently has the following members in it. This list is always sorted in descending order.

Name	Rank
Zachary	1
Samuel	2
Rudolph	3
Roshan	4
Brandi	5
Anita	6

Rank the list again after adding the following members to the list.
    Sandy, Rudy, Brendon, Anderson

Name	Rank
Zachary	1
Sandy	2
Samuel	3
Rudy	4
Rudolph	5
Roshan	6
Brendon	7
Brandi	8
Anita	9
Anderson	10

Are the rankings of all the members the same now? No. Note that Sandy is listed before Samuel because "n" comes before "m" in descending order. Verify if the list is in ascending order from bottom to top.

Answers

1 SEQUENCING

February						
Sunday	Monday	Tuesday	Wednesday	Thursday	Friday	Saturday
				1	2	3
4	5	6	7	8	9	10
11	12	13	14	15	16	17
18	19	20	21	22	23	24
25	26	27	28			

Raja takes piano lessons every thursday.
1) When was his last class in February? Answer: B) February 22
2) Which of the following is the thursday after his last class?
        Answer: B) March 1

Joshua takes lessons in gymnastics every Monday, Wednesday, and Friday.

1) How many gymnastics lessons did he attend in the first week of February? Answer: C) 1 lesson (Feb 2)

2) How many gymnastics lessons did he attend in the last week of February?   Answer: B) 2 lessons   (Feb 26, 28)

3) For how many weeks did he take the lessons three times each week? Answer: 3 (weeks of Feb 4, Feb 11 and Feb 18)

4) How many gymnastic lessons did Joshua take in February? Answer: A) 12 (Feb 2,5,7,9,12,14,16,19,21,23,26,28)

Answers

© Gift Of Logic, Inc * Copying prohibited

1 SCHEDULING

Doctor Harper's Appointment Schedule

	Feb 21	Feb 22	Feb 23
8 AM	Jenny		Arjun
9 AM		Mohan	
10 AM	Asif		Josh
11 AM		Laura	Hana

1) If Jenny wants to change her appointment to the another time on the same day, how many choices does she have?
    Answer: B) 2 - either 9 AM or 11 AM

2) If Asif wants to change his appointment to another day at the same time, how many choices does he have?
    Answer: A) 1 - on Feb 22

3) If Mohan wants to change his appointment to another day at another time, how many choices does he have?
    Answer A) 1 - He can change his appointment to Feb 21 at 11 AM. He cannot change to another time on Feb 22. He also cannot change to 9 AM on Feb 21 or Feb 23.

4) If Laura wants to change her appointment to another day at a time that is not 9 AM, how many choices does she have?
    Answer: A) 1 - on Feb 21 at 11 AM

5) Hana wants her appointment to be rescheduled to 8 AM. On which day can she now see the doctor?
    Answer: B) Feb 22. Feb 21 and Feb 23 are already booked.

Answers

© Gift Of Logic, Inc * Copying prohibited

1 LOOPING

Looping means repeating several times. The looping stops when some condition is met. You keep track of loops in a counter.

Write your name in the column marked 'Name'. Every time you write your name, increment (add) the counter by one. Keep writing your name until the counter shows 8.

Name	Counter
Emma	1
Emma	2
Emma	3
Emma	4
Emma	5
Emma	6
Emma	7
Emma	8
Emma	9
Emma	10

1) What does the counter keep track of?
Number of times the name is written.

2) If you write your name two more times, what number will the counter show? 10

3) If the counter runs up to 15, how many times would you have written your name? 15

Answers
© Gift Of Logic, Inc * Copying prohibited

2 LOOPING

Christina bought a new piggy bank. Her dad said he will keep on dropping a dime (10 cents) in it until it adds up to one dollar (100 cents).

Counter	Coins	Total
1	10 cents	10 cents
2	10 cents	20 cents
3	10 cents	30 cents
4	10 cents	40 cents
5	10 cents	50 cents
6	10 cents	60 cents
7	10 cents	70 cents
8	10 cents	80 cents
9	10 cents	90 cents
10	10 cents	100 cents

Fill in the grid - write 10 cents in the 'coins' column every time a dime is dropped into the piggy bank. Keep track of how many times a dime was dropped in the column marked 'Counter'. Keep track of the Total as well.

1) When the counter reads 6, how many dimes were dropped? Ans: 6
2) When the 'Total' column reads 40 cents, how many dimes were dropped?   Ans: 4
3) When the 'Total' column reads 100 cents, how many times were the dimes dropped? Ans: 10
4) What is the counter reading when 80 cents worth of coins were dropped? Ans: 8
5) Can this counter go beyond 10?
      Answer: B) No. Christina's dad will drop coins only until it adds up to 100 cents-that is, until the counter is 10.

Answers
© Gift Of Logic, Inc * Copying prohibited

1
FIFO (FIRST IN, FIRST OUT)

The Genius Elementary School invited a face painter..

1) Who was the first one to have his face painted?
   Answer: B) Dan - since he was the first one in the line.

2) Who was the last one to have his face painted?
   Answer: C) David - since he was the last one on the line.

3) The last person in the line is the last one to have his face painted.
   Answer: A) True - the first one in the line gets his face painted first and the last one in the line gets his face painted last.

Tony first arrived at the doctor's office..
  The order of arrival is Tony, Andy, Gina

1) Who did the doctor see first?
   Answer: A) Tony

2) Did the doctor see Gina before Andy?
   Answer: B) No
   Andy came before Gina and will be seen before Gina.

3) Who did the doctor see last?
   Answer: A) Gina

4) Write the order in which the doctor saw the patients.
   Tony, Andy, Gina

Answers

© Gift Of Logic, Inc * Copying prohibited

2 LIFO (LAST IN, FIRST OUT)

Last In First Out is the same as First In Last Out

Name	Row
Rick	3
Martin	4
Chuck	3
William	1
Nancy	2
Debby	1
Drew	2

1) Martin sat in the last row and was the last one to leave the conference.
Answer: B) False - he was the first to leave the conference.

2) Nancy left the conference before Rick did.
Answer   B) False. Nancy was in row# 2 and Rick in row# 3. So, Rick left before Nancy.

3) Debby and William left the conference after Nancy and Drew.
Answer: A) True. They sat in the first row and left after Nancy and Drew who sat in the second row.

4) Who were the last people to leave the conference?
Answer: William and Debby sat in the first row and were the last ones to leave.

answers

CORRELATION

1

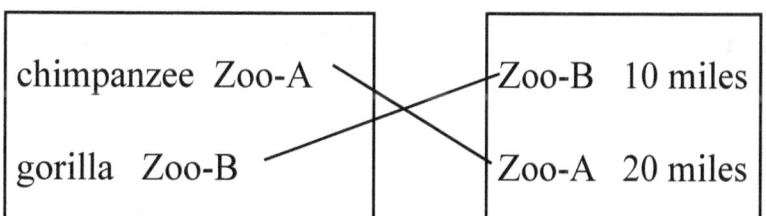

1) How far should we go to see the gorilla? 10 miles
2) How far should we go to see the chimpanzee? 20 miles

---

2 Several students are playing the school playground. The following information is available. Using this information, answer the questions below.

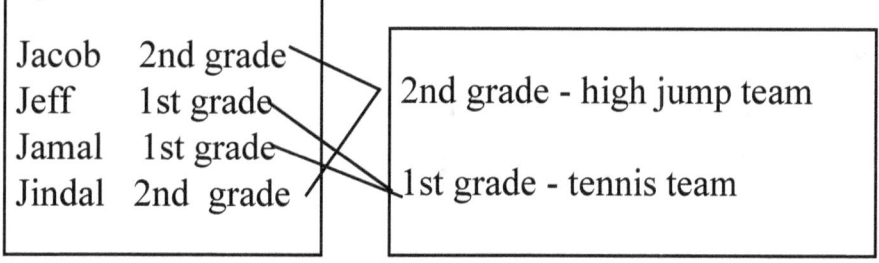

1) In which team can we find Jamal in? Answer: tennis team

2) In which team can we find Jindal in? Answer: high jump team

3) Who are all in the high jump team? Jacob, Jindal

4) Who are all in the tennis team? Jeff, Jamal

Answers

GROUPING

1

All the city buses are painted blue. Bus# 22 is a city bus.

What color is Bus #22? Answer: Blue
Why? Bus# 22 is blue because it is a city bus and all city buses are painted blue.

2

music club - at least one instrument
drama club - acting skills

Calvin can play one instrument.
Trevor can play two instruments as well as act.
Josh can neither play an instrument nor act.

Based on the information given above, answer the following questions.

1) Who is eligible to become a member of the Music club?
    Calvin, Trevor

2) Who is eligible to become a member of the Drama club?
    Trevor

3) Which clubs can Josh become a member of?
    None, since he can neither play an instrument nor act.

Answers
© Gift Of Logic, Inc * Copying prohibited

GROUPING

3 The following are the names of the players in the Alpha Sports club..

1) How many members are there in the sports club? 10

2) Who are all members of the soccer group and the baseball group? Write their names in the list below.

Soccer Group	Baseball group
Randy	Randy
William	Mark
Gary	Gary
Zac	Sean
Anita	Britney
Priti	Priti
Sidney	

3) How many members are there in the soccer group? 7

4) How many members are there in the baseball group? 6

answers

GROUPING (continued)

4 5) Add up the number of members in the soccer and baseball group. Is this number the same as the number of members in the sports club? Explain.

No, there are 7 members in the Soccer group and 6 in the baseball group giving us a total of 13 members in both groups. But, the sports club has only 10 members. This is because there are some members who belong to both groups.

6) Who are all members of both groups? Randy, Gary, Priti

Now, fill in the three groups below with the names of players from the Alpha Sports Club.

Soccer Group	Soccer and Baseball Group	Baseball Group
William Zac Anita Sidney	Randy Gary Priti	Mark Shawn Britney

7) How many members are there in the soccer group? 4

8) How many members are there in the baseball group? 3

9) How many members are there in the soccer and baseball group? 3

10) Add up the number of members in the soccer, soccer and baseball, and baseball groups. Is this number the same as the number of members in the sports club?   Yes (10 members)

Answers
© Gift Of Logic, Inc * Copying prohibited

GROUPING AND SUMMARIZING

5 The residents of a Zoo and their population are shown below.

Residents	Population
Lions	2
Parrots	5
Tigers	3
Eagles	3
Hippopotamus	4
Flamingoes	3

Split the group into birds and animals and write the names of the members of the group and their population in the two boxes shown below.

Birds
Parrots -5
Eagles -3
Flamingoes -3

Animals
Lions -2
Tigers -3
Hippopotamus - 4

1) How many animals are there in the Zoo? 9
2) How many birds are in the Zoo? 11
3) There are more animals in the Zoo than there are birds.
   Answer: B) False - there are 9 animals and 11 birds in the Zoo.
4) More birds must be brought in so that there are equal number of animals and birds in the Zoo.
   Answer: B) False. There are already more birds than animals.

Answers
© Gift Of Logic, Inc * Copying prohibited

1 VENN DIAGRAM

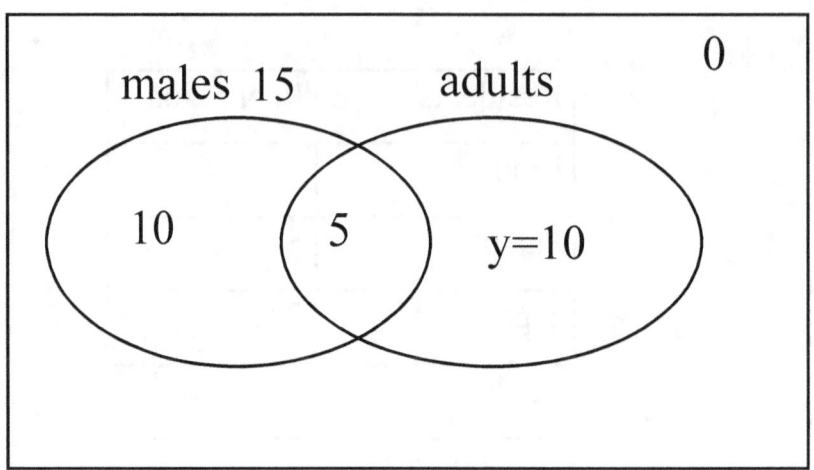

There are 10 males in the room who are not adults. There are 5 adult males in the room. Everyone in the room is either a male or an adult. There are totally 25 people in the room.

Complete the Venn diagram and find the number of adults who are not male.

$$10 + 5 + y = 25$$
$$y = 10$$

This is shown in the diagram. These are the adults that are not male.

1) How many adults in the room are not male?
    10 (as discussed above)

2) How many adults are there in the room?
    5 (male adults) + 10 (adults that are not male) = 15 adults

3) How many males are there in the room?
    10 (males who are not adults) + 5 (male adults) = 15 males

Answers
© Gift Of Logic, Inc * Copying prohibited

2 VENN DIAGRAM

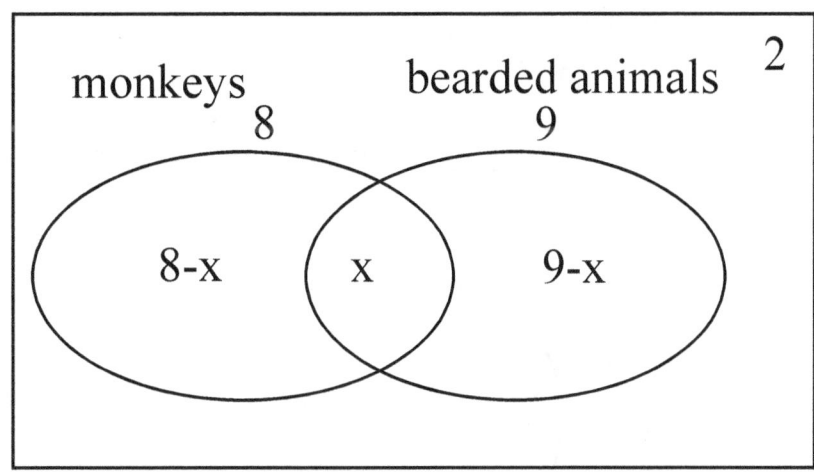

20 animals in a Zoo, 8 are monkeys
there are 9 animals with beards
2 animals are neither monkeys nor do they have beards

First, identify the two groups as monkeys and bearded animals. Complete the Venn diagram as shown above and calculate x.
   $8-x+x+9-x+2=20$; So, $x = 1$

1) How many monkeys have beards?
   $x=1$ monkey.

2) How many monkeys do not have beards?
   $8-x = 8-1 = 7$ monkeys

3) How many animals with beards are not monkeys?
   $9-x = 9-1 = 8$ animals

4) How many animals are either monkeys or bearded?
   $8-x+x+9-x = 17-x = 17-1 = 16$ animals

Answers
© Gift Of Logic, Inc * Copying prohibited

3 VENN DIAGRAM

Group A: Monkeys   Group B: Animals   Group C: Animals that live in water.

A) Incorrect - All monkeys are animals - so circle A must be completely inside circle B.

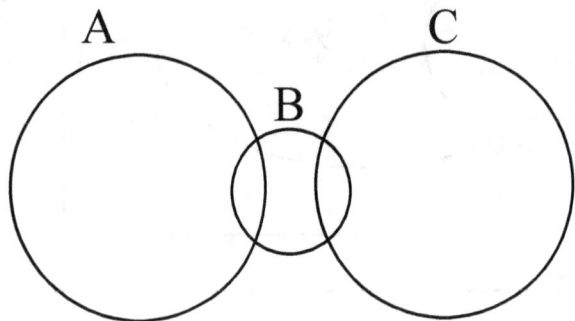

B) Incorrect - relation between groups A and C is incorrect in this diagram. Since monkeys do not live in water, circles A and C must not intersect.

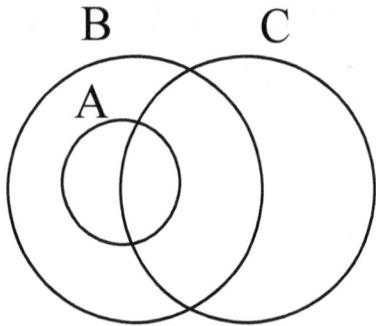

C) Correct - All animals are monkeys. No monkey lives in water. Some animals live in water (like hippopotamus).

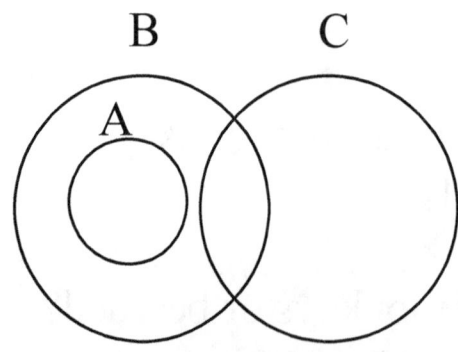

Answers
© Gift Of Logic, Inc * Copying prohibited

1 GRAPH LOGIC

The bar graph above shows the number of apples that Patrick ate during three years.

1) Patrick ate more apples each year compared to its previous year.
Answer: B) False. In 2007, Patrick ate less number of apples than he did in 2006.

2) If Patrick had eaten six more apples in 2007, then 2007 would be the year when he ate the most apples. Answer: A) True.
If he had eaten 6 more apples in 2007, he would have eaten 35+6=41 apples. This is more than 25 and 40, the apples that he ate in other years.

2 GRAPH LOGIC

The clustered bar graph above shows the head count of the number of boys and girls in a day care facility.

1) There were more boys in 2007 than in 2006.
Answer: A) True. There were 10 boys in 2006 and 30 in 2007.

2) There were less number of girls in 2007 than in 2006.
Answer: B) False. There were more number of girls in 2007 than in 2006.

Answers
© Gift Of Logic, Inc * Copying prohibited

3 GRAPH LOGIC

The clustered bar graph shows the number of soccer and baseball games played in a park in two different years.

1) More soccer games were played in 2006 than in 2005.
Answer: A) True. The bar chart for soccer is longer in 2006 than in 2005.

2) Less number of baseball games were played in 2005 than in 2006
Answer: B) False. The bar chart for baseball is longer in 2005 than in 2006.

3) More games were played at the park in 2005 than in 2006.
Answer: B) False. Totally 30 games were played in the park during both years.

4 GRAPH LOGIC

The line graph above shows the value of a car as it ages.
1) As a car becomes older, its value decreases.
Answer: A) True. This can be seen clearly from the slope of the graph.

2) Newer a car, the more expensive it is.
Answer: B) True. When the car is 1 or 2 years old, it is more expensive than when it is 3 or 4 years old. Again, the slope of the graph shows this fact clearly.

3) The slump in value from 15,000 to 10,000 dollars started when the car was three years old.
Answer: B) False. The slump started when the car was two years old.

Answers
© Gift Of Logic, Inc * Copying prohibited

5 GRAPH LOGIC

The line graph shows the water level in a tank..

1) Water level was constant between 8 AM and 9 AM. Answer: B) False. The water level fell from 10 feet to 5 feet from 8 AM to 9 AM.

2) Sometime between 9 AM and 10 AM, the water level in the tank was 41 feet. Answer: B) False. The maximum water level between 9 AM and 10 AM was 40 feet.

3) The water level in the tank dropped by the same amount from 8 AM to 9 AM and from 10 AM to 11 AM. Answer: B) False. Although the water level fell from 8 AM to 9 AM and from 10 AM to 11 AM, it fell by 5 feet and 10 feet respectively.

6 GRAPH LOGIC

The pie-graph shows the amount of time spent by Jane..

1) The predominant activity during the summer was
Answer: B) sleeping

2) Compared to sports, less time was spent on reading.
Answer: A) True. 25% of time was spent on reading and 35% on sports.

3) Jane slept more time than playing sports and reading books combined.
Answer: B) False. Jane spent 60% of her time on sports and reading combined, but only 40% on sleeping.

Answers

7
GRAPH LOGIC

The temperature from 10 AM to Noon in two cities A and B are shown in the line graph.

1) Between 10 and 11 in the morning, city B is cooler than city A.
Answer: B) False. City A is cooler than city B from 10 to 11 AM.

2) City A is always cooler than city B.
Answer: B) False. After 11 AM, city B is cooler than city A.

8
GRAPH LOGIC

The graph above shows the level of water in a water tank ..

1) Water was not used from the tank between 9 AM and 10 AM.
Answer: A) True. Even though water could be used from the tank from 9 AM, it was not used until 10 AM. The graph is flat between 9 AM and 10 AM.

2) Water was drained from the tank only from 10 AM to Noon.
Answer: A) True. Water was filled from 8-9 AM and remained constant from 9-10 AM. From 10 AM to Noon, it was used and the graph clearly shows that the level went down during this time.

Answers
© Gift Of Logic, Inc * Copying prohibited

NUMBER LOGIC

Figure out the logic in the sequence and find the missing number.

1

½   1   1 ½   ?   Answer: 2 - numbers increase by ½

2

2 ½   2   1 ½   ?   Answer: 1 - numbers decrease by ½

3

¼   ½   ¾   ?   Answer: 1 - numbers increase by ¼

4

4   8   16   ?   Answer: 32 - numbers multiply by 2

5

3   6   ?   24   Answer: 12 - each number is multiplied by 2

6

2   1   ½   ?   Answer: ¼ - each number divided by 2 gives the next

---

7

10   5   ?   1 ¼   Answer: 2 ½ - each number is divided by 2 to get the next number

---

8

4   1   ?   1/16   Answer: ¼ - each number divided by 4 gives the next number

---

Answers

© Gift Of Logic, Inc * Copying prohibited

NUMBER LOGIC

Figure out the logic in the numbers and find the missing number.

9 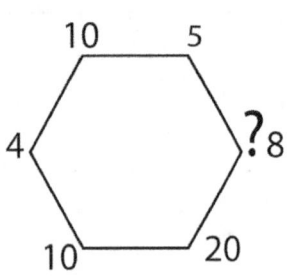 Numbers that are across each other are doubled. Number across 4= 4*2=8.

10 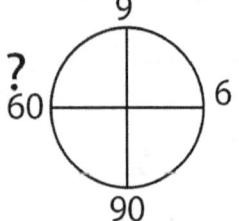 9*10=90. 6*10=60.

11 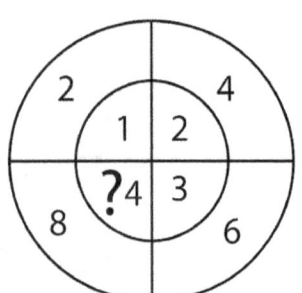 Numbers in the outer segment are multiples of the numbers in the inner segment. 1*2=2, 2*2=4, 3*2=6 and so 4*2=8.

12 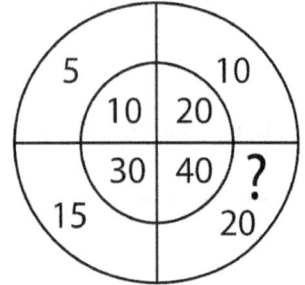 Numbers in the outer segment are half the numbers on the inner segment. 10/2=5, 20/2=10, 30/2=15, 40/2=20.

13 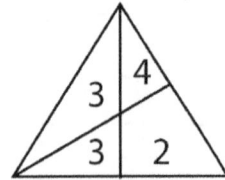 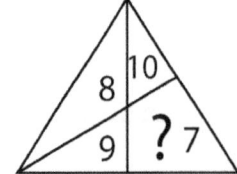 Numbers across increase by 1 from bottom to top.

Answers

© Gift Of Logic, Inc * Copying prohibited

NUMBER LOGIC

Figure out the logic in the numbers and find the missing number.

14 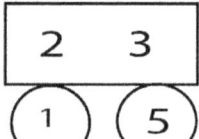 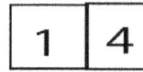   6+3=9; 4+1=5

15   3-2=1; 3+2=5; 6-5=1; 6+5=11

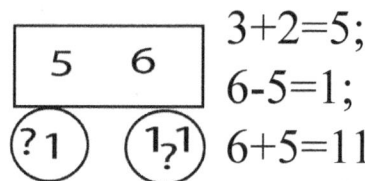

16   add the numbers and write it at the top. 6+2+4+1+9+3=25

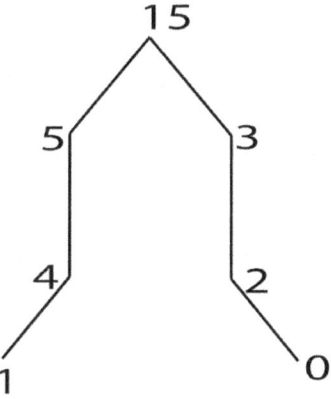

17   4+3=7; 4*3=12
6+3=9; 6*3=18

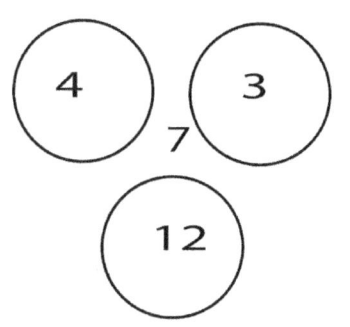

18   divide by 10

100,000   10,000   ? 1000   100

Answers   127
© Gift Of Logic, Inc * Copying prohibited

LETTER LOGIC

Figure out the logic in the sequence and find the missing letter or number.

1
A2 B4 C? D8   Answer: 6 - A,B,C,D and 2,4,6,8

2
ACE MOQ V?? Answer: X Z - skip an alphabet to get the next

3
AZ BY C? Answer: X - pick one letter from beginning and one from end

4
C1E K2M Q?S Answer: 3 - Numbers in the middle increase in sequence

5
AUG OCT ? Answer: DEC - first three letters of the month- skip one

6
J F M A ? Answer: M - first letters of the month - J for January, F for February. M for May

7
SUN TUE THU ? Ans: SAT - first three letters of the day - skip one. SAT=first three letters of Saturday

Answers
© Gift Of Logic, Inc * Copying prohibited

LETTER LOGIC   LHS=left hand side   RHS=right hand side

Figure out the logic in the sequence and find the missing letter or number.

8	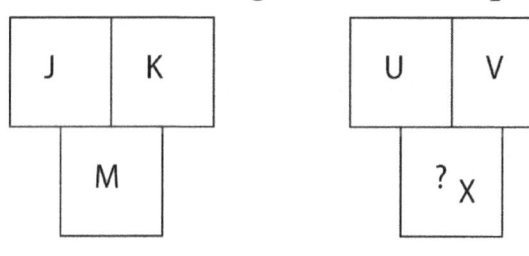	On LHS, J, K, skip one and then M. On RHS, U,V, skip W and you get X.
9	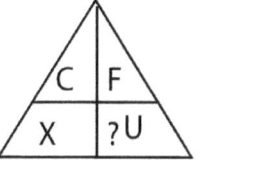	On LHS, C is the 3rd letter from beginning and X is the 3rd from the end. On RHS, F is the 6th letter from beginning and U is the 6th letter from the end.
10	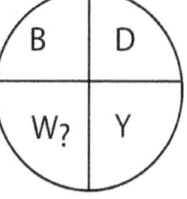	From B, skip one letter, you get D. W, skip one and then Y.
11	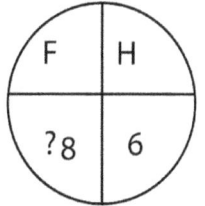	F is the 6 the letter, it is opposite to F. H is the 8th letter.
12	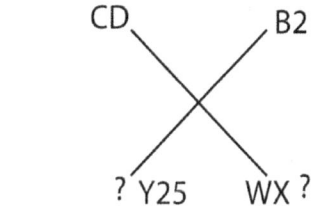	On LHS, A1 and Z26 form a logical pair. Also AB and YZ are 2 letter pairs from beginning and end. On RHS, B2 will correspond to Y25, the second last letter with its associated number. Similarly CD and WX are the 2 letters from the beginning and end of the alphabetic sequence.

Answers

© Gift Of Logic, Inc * Copying prohibited

LETTER LOGIC

Figure out the logic in the sequence and find the missing letter or number.

13

MNP   QRT   ?   Answer: UVX - Two letters in sequence, skip one.

14

LmN   OpQ   ? Answer: RsT - Letter in middle is lower case.

15

A4C   E16K   I?Q   Answer: 26 - A is 1, C is 3, 1+3=4, so, A4C. E is 5, K is 11, So, E5+11K= E16K, Similarly, I is 9, Q is 17, I9+17Q= I26Q.

16

A1B2   X24?26   Answer: Z - A is 1, B is 2 in alpha sequence. X is 24, Z is 26.

17

    Az   By   ?   Answer: Cx - Capital letters go from A to C. Small letters go in reverse from z to x.

18

    aAz   bBy   ?   Answer: cCx - Capital letters in the middle, small letter in the left, small letter in reverse order in the right.

---

19 A^B B^C ? Answer: C^D
letters in pairs, one letter is raised a bit.

Answers

1
SUDOKU

Solve the following Sudoku. A correctly solved Sudoku has numbers 1-9 appearing only once in each row, each column and each 3x3 grid. You gain valuable positioning skills by solving these sudokus.

4	1	8	2	9	6	5	3	7
3	5	6	7	8	4	9	2	1
9	2	7	3	5	1	6	8	4
2	3	4	5	6	9	7	1	8
6	7	1	4	3	8	2	5	9
8	9	5	1	2	7	3	4	6
7	8	2	9	4	5	1	6	3
5	4	9	6	1	3	8	7	2
1	6	3	8	7	2	4	9	5

Answers

© Gift Of Logic, Inc * Copying prohibited

2 SUDOKU

Solve the following Sudoku. A correctly solved Sudoku has numbers 1-9 appearing only once in each row, each column and each 3x3 grid. You gain valuable positioning skills by solving these sudokus.

2	9	3	5	6	8	1	7	4
5	8	7	4	9	1	2	6	3
6	4	1	7	3	2	5	9	8
8	5	4	6	1	9	7	3	2
3	7	2	8	4	5	6	1	9
1	6	9	2	7	3	4	8	5
9	2	5	1	8	7	3	4	6
4	1	8	3	5	6	9	2	7
7	3	6	9	2	4	8	5	1

3 SUDOKU

Solve the following Sudoku. A correctly solved Sudoku has numbers 1-9 appearing only once in each row, each column and each 3x3 grid. You gain valuable positioning skills by solving these sudokus.

4	7	9	1	2	8	6	3	5
5	2	3	9	4	6	8	7	1
8	6	1	5	7	3	2	9	4
3	8	4	2	5	1	9	6	7
9	5	2	7	6	4	3	1	8
6	1	7	3	8	9	5	4	2
2	4	5	6	9	7	1	8	3
1	9	8	4	3	2	7	5	6
7	3	6	8	1	5	4	2	9

4 SUDOKU

Solve the following Sudoku. A correctly solved Sudoku has numbers 1-9 appearing only once in each row, each column and each 3x3 grid. You gain valuable positioning skills by solving these sudokus.

8	9	2	7	5	3	4	6	1
1	7	6	2	4	9	3	8	5
3	5	4	8	1	6	9	7	2
4	6	8	5	9	2	7	1	3
9	2	5	3	7	1	6	4	8
7	3	1	6	8	4	2	5	9
5	4	7	9	3	8	1	2	6
2	1	9	4	6	5	8	3	7
6	8	3	1	2	7	5	9	4

Answers
© Gift Of Logic, Inc * Copying prohibited

5

SUDOKU

Solve the following Sudoku. A correctly solved Sudoku has numbers 1-9 appearing only once in each row, each column and each 3x3 grid. You gain valuable positioning skills by solving these sudokus.

8	5	7	1	6	4	2	3	9
9	1	6	2	3	8	7	5	4
3	2	4	9	5	7	1	8	6
4	8	9	7	1	2	3	6	5
5	7	1	3	8	6	4	9	2
2	6	3	5	4	9	8	1	7
1	9	8	4	7	5	6	2	3
7	3	2	6	9	1	5	4	8
6	4	5	8	2	3	9	7	1

Answers

6 SUDOKU

Solve the following Sudoku. A correctly solved Sudoku has numbers 1-9 appearing only once in each row, each column and each 3x3 grid. You gain valuable positioning skills by solving these sudokus.

3	4	8	1	9	5	7	2	6
5	2	9	4	7	6	1	8	3
1	7	6	2	3	8	5	4	9
7	9	1	8	2	3	6	5	4
6	8	3	5	4	7	2	9	1
4	5	2	9	6	1	3	7	8
9	6	4	3	5	2	8	1	7
2	1	7	6	8	9	4	3	5
8	3	5	7	1	4	9	6	2

Answers
© Gift Of Logic, Inc * Copying prohibited

PICTURE SEQUENCE

Figure out the logic in the picture sequence, and draw the next picture in the sequence.

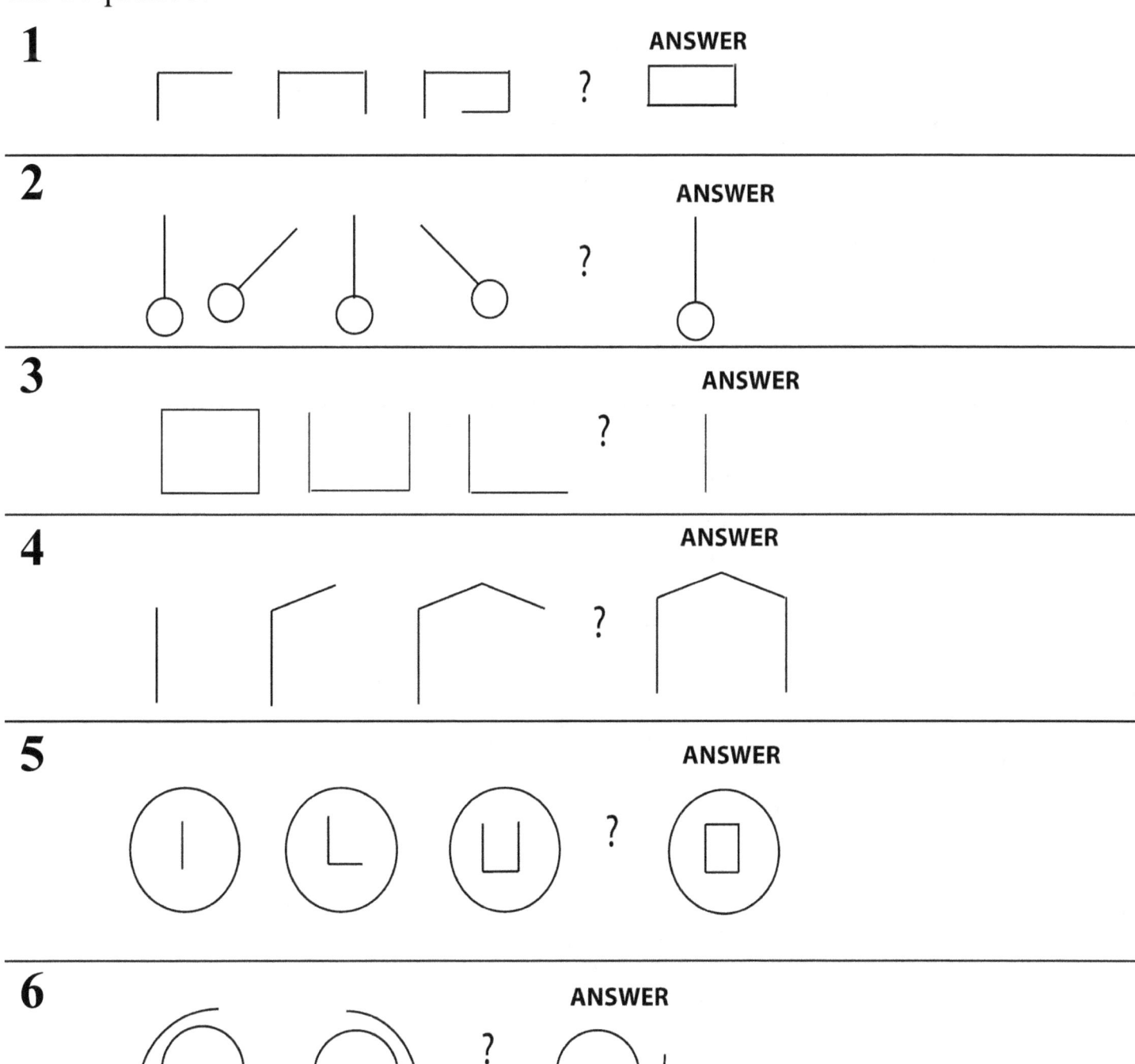

PICTURE SEQUENCE

Figure out the logic in the picture sequence, and draw the next picture in the sequence.

7 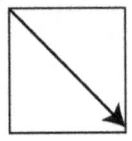 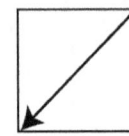 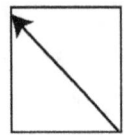 ? **ANSWER**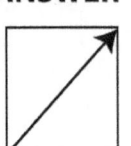

8 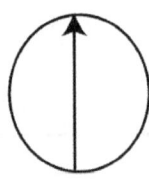 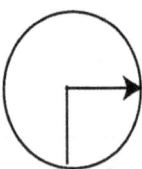 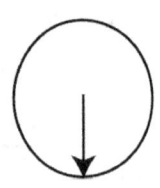 ? **ANSWER**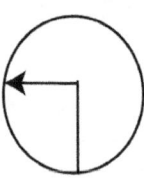

9 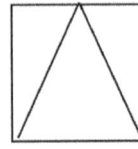 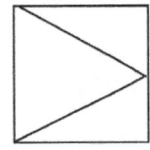 ? **ANSWER**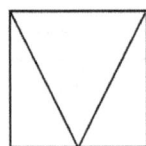

10   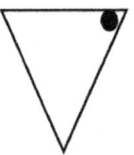 ? **ANSWER**

11 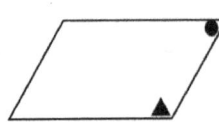 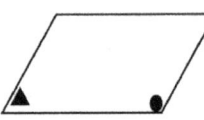  ? **ANSWER**

Answers

PICTURE SEQUENCE

Figure out the logic in the picture sequence, and draw the next picture in the sequence.

12

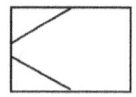 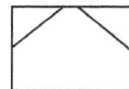  ? **ANSWER**

13

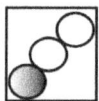 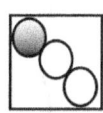 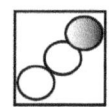 ? **ANSWER**

14

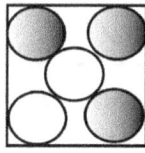 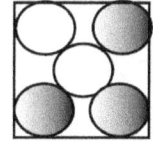 ? **ANSWER**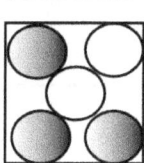

15

   ? **ANSWER**

16

 ? **ANSWER**

Answers 139

© Gift Of Logic, Inc * Copying prohibited

PICTURE ANALOGY

Figure out the logic in the picture analogy, and circle the correct picture.

1 □ : □ AS □ : □ □ — **B ANSWER**

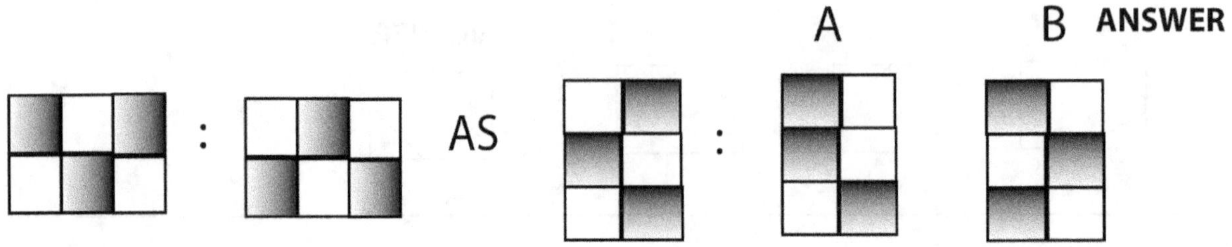

2 ○ : ○ AS ○ : ○ ○ — **B ANSWER**

3 ⊥ : ⊥ AS ✕ : ✕ ✕ — **A ANSWER**

4 △ : △ AS ▭ : ▭ ▭ ▭ — **C ANSWER**

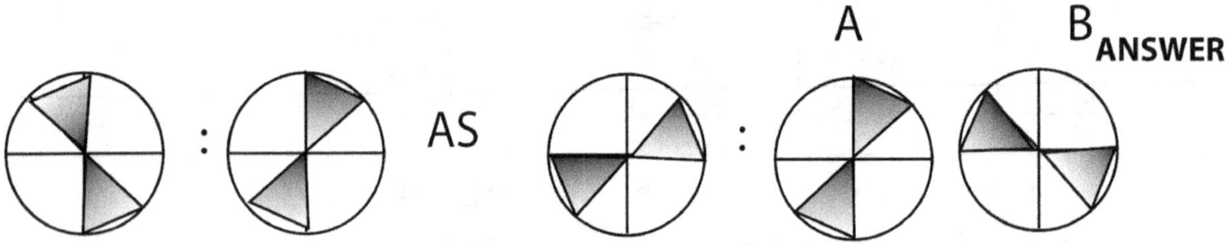

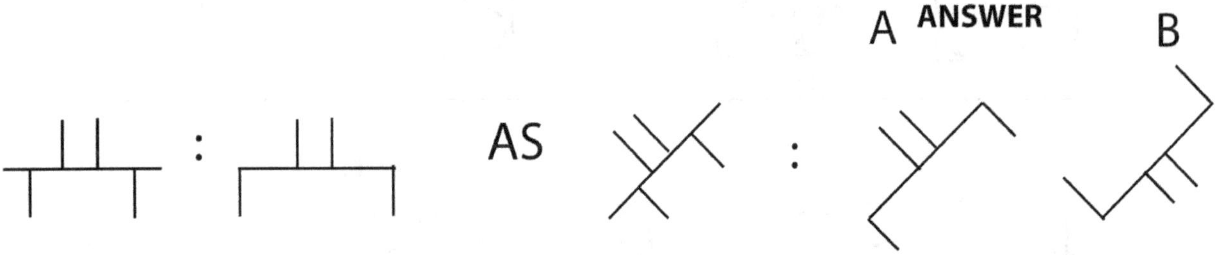

Answers

140

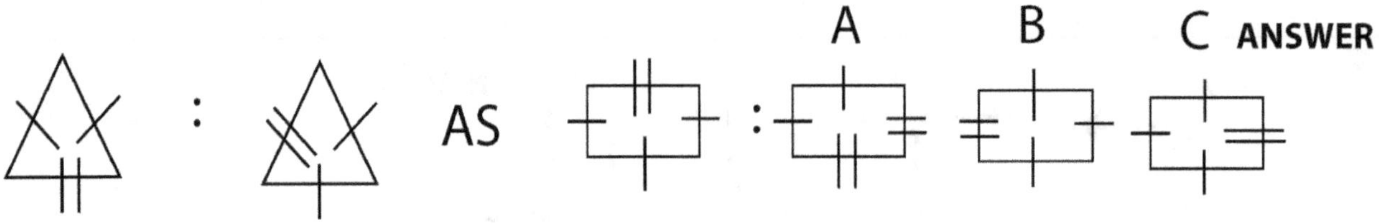

© Gift Of Logic, Inc * Copying prohibited

PICTURE ANALOGY

Figure out the logic in the picture analogy, and circle the correct picture.

5 ○ : ◎ AS □ : (A) overlapping rectangles (B) rectangle with small square (C ANSWER) rectangle within rectangle

6 △ : cone AS dome-rectangle : (A) rectangle with rounded end (B ANSWER) cylinder shape (C) rectangle with curved bottom

7 ⌒ : ⌒ AS ○ : (A ANSWER)) ((B) ( )

8 □ : ▭ AS ⬡ : (A ANSWER) short trapezoid (B) tall trapezoid

Answers
© Gift Of Logic, Inc * Copying prohibited

PICTURE ANALOGY

Figure out the logic in the picture analogy, and circle the correct picture.

9 A   B ANSWER

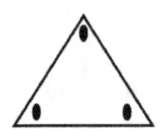 : 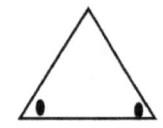 AS 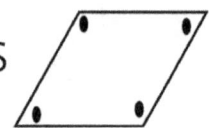 :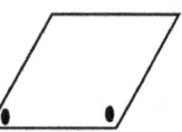

10 A   B   C ANSWER

 : AS 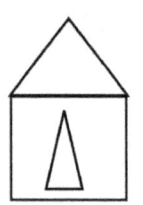 :

11 A   B   C ANSWER

 : AS  :

12 A   B ANSWER

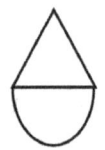 :  AS 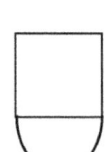 :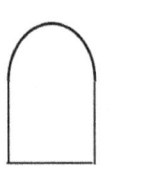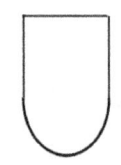

13 A   B   C ANSWER

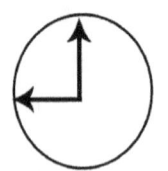 : AS 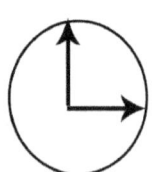 :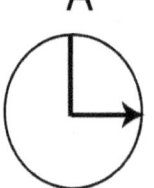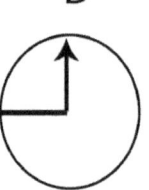

Answers
© Gift Of Logic, Inc * Copying prohibited

ODD PICTURE

Figure out the logic in the pictures, and identify the odd picture.

1   A   B ANSWER   C

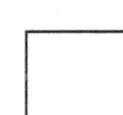

four sides in B

---

2   A   B ANSWER   C

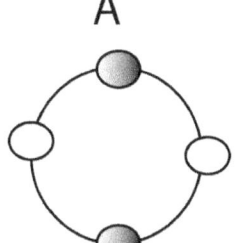

shaded ovals are not opposite to each other in B

---

3   A   B   C ANSWER

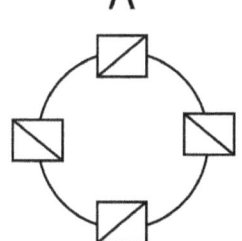

diagonal line through the opposite square pairs is not in the same direction in C

---

4   A   B ANSWER   C

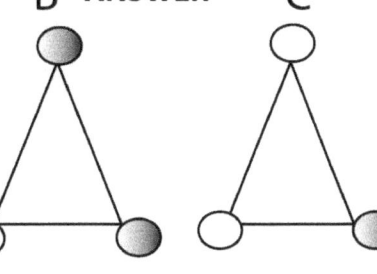

two ovals are shaded in B

---

5   A   B   C ANSWER

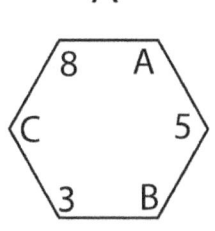 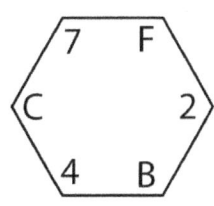 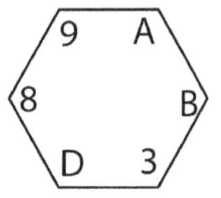

two consecutive letters in C

Answers
© Gift Of Logic, Inc * Copying prohibited

ODD PICTURE

Figure out the logic in the pictures, and identify the odd picture.

6   A   B   C **ANSWER**

 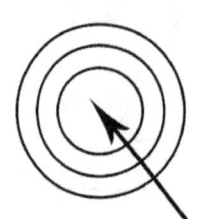 

arrow is not pointing at the center in C

7   A   B   C **ANSWER**

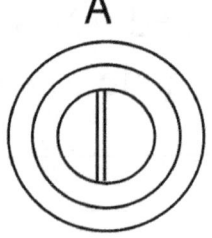 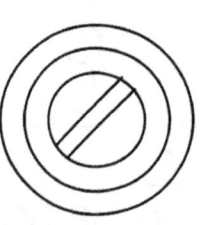 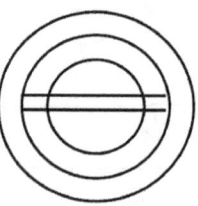

double line is not within the inner circle in C

8   A   B **ANSWER**   C

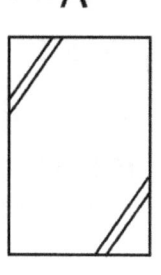 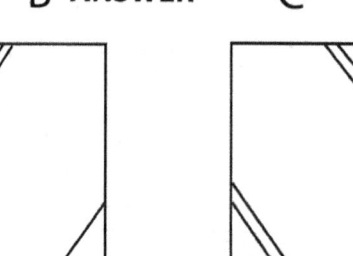

one line missing in B

9   A **ANSWER**   B   C

 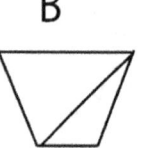 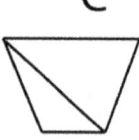

line is not from edge to edge in A

10   A   B **ANSWER**   C

shapes do not alternate in B

Answers

144

© Gift Of Logic, Inc * Copying prohibited

PICTURE DIFFERENCE

Mark the differences in the set of pictures shown, with arrows.

1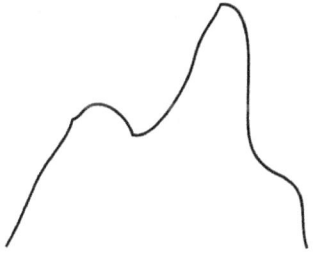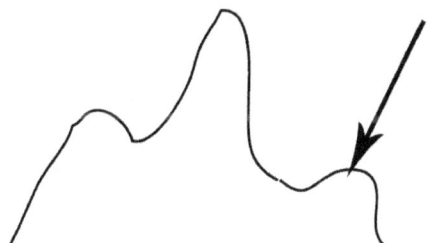

2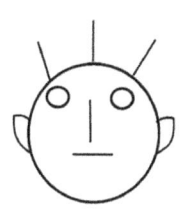

3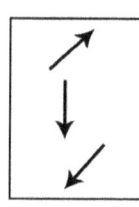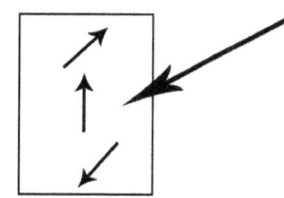

4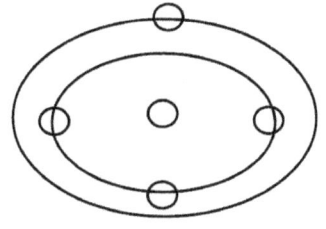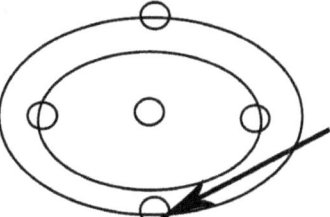

5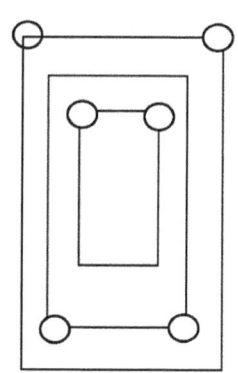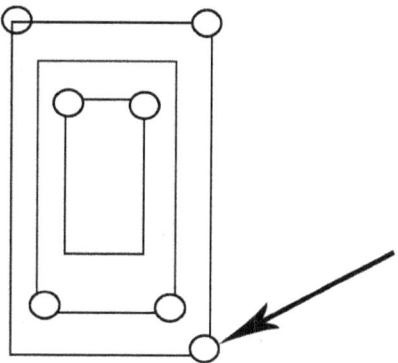

Answers

© Gift Of Logic, Inc * Copying prohibited

PICTURE DIFFERENCE

Mark the differences in the set of pictures shown, with arrows.

6

7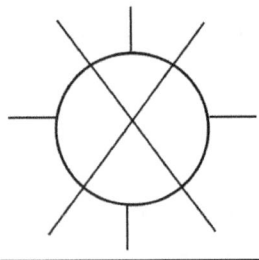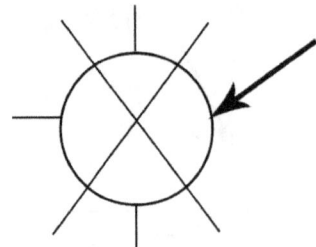

8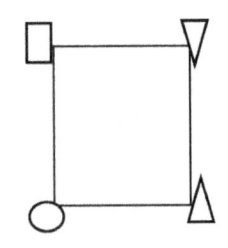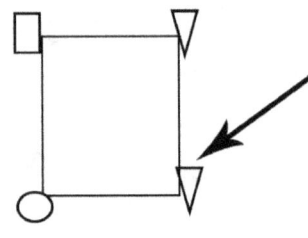

9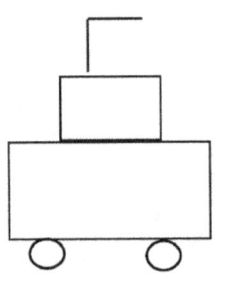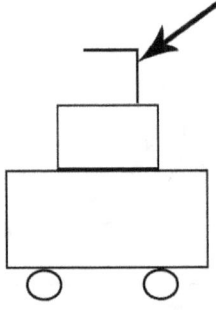

10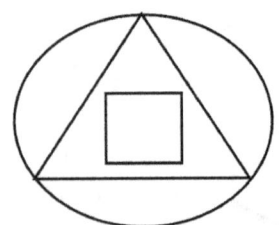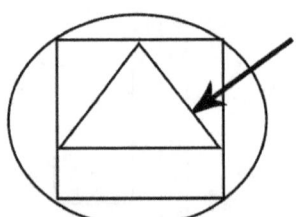

Answers
© Gift Of Logic, Inc * Copying prohibited

PICTURE DIFFERENCE

Mark the differences in the set of pictures shown, with arrows.

11

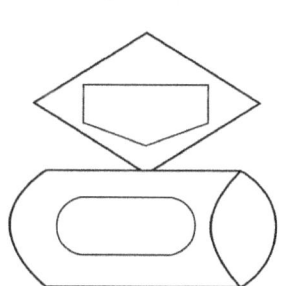 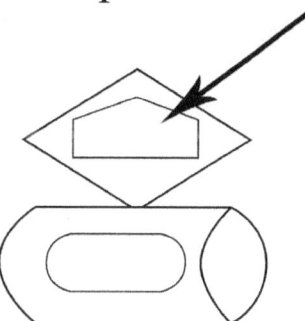

12

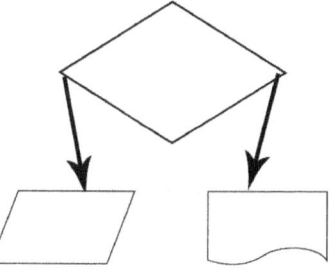 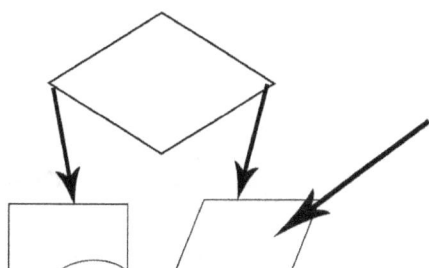

13

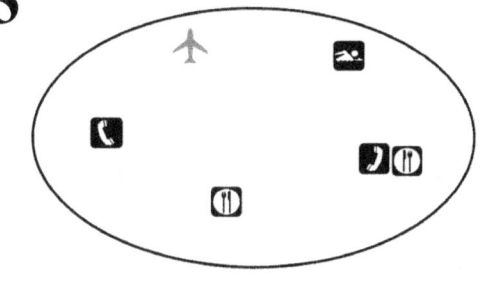 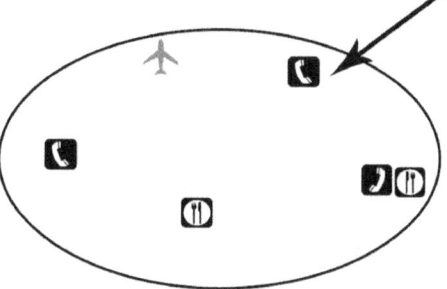

14

15

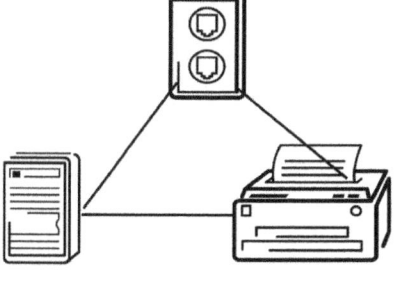 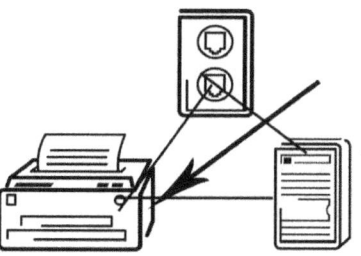

Answers

PATTERN MATCHING

Find the logical pattern in the pictures on the left, and identify the picture on the right that will fit in the space marked with ? to complete the pattern.

1   A   B ANSWER   C

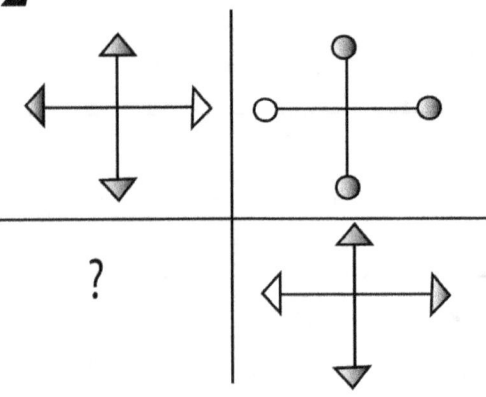 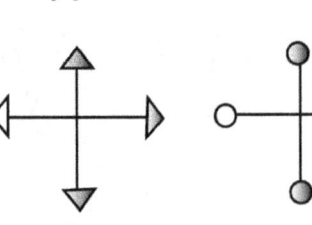 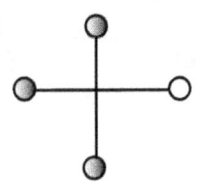

2

              A   B   C ANSWER

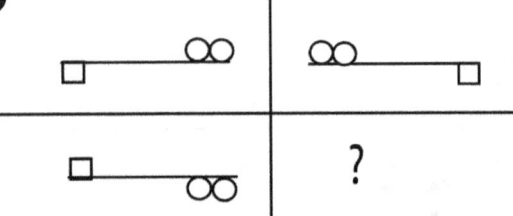

3

              A   B   C ANSWER

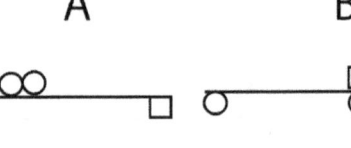

4   A   B ANSWER   C

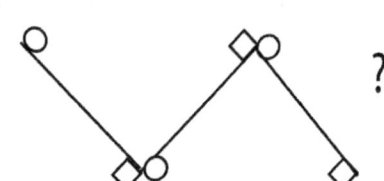 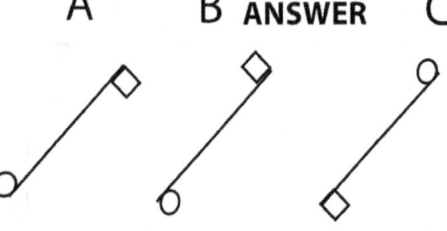

Answers

148

© Gift Of Logic, Inc * Copying prohibited

NOTES